DAMS:
IMPACTS, STABILITY AND DESIGN

DAMS:
IMPACTS, STABILITY AND DESIGN

WALTER P. HAYES
AND
MICHAEL C. BARNES
EDITORS

Nova Science Publishers, Inc.
New York

For permission to use material from this book please contact us:
Telephone 631-231-7269; Fax 631-231-8175
Web Site: http://www.novapublishers.com

NOTICE TO THE READER

The Publisher has taken reasonable care in the preparation of this book, but makes no expressed or implied warranty of any kind and assumes no responsibility for any errors or omissions. No liability is assumed for incidental or consequential damages in connection with or arising out of information contained in this book. The Publisher shall not be liable for any special, consequential, or exemplary damages resulting, in whole or in part, from the readers' use of, or reliance upon, this material. Any parts of this book based on government reports are so indicated and copyright is claimed for those parts to the extent applicable to compilations of such works.

Independent verification should be sought for any data, advice or recommendations contained in this book. In addition, no responsibility is assumed by the publisher for any injury and/or damage to persons or property arising from any methods, products, instructions, ideas or otherwise contained in this publication.

This publication is designed to provide accurate and authoritative information with regard to the subject matter covered herein. It is sold with the clear understanding that the Publisher is not engaged in rendering legal or any other professional services. If legal or any other expert assistance is required, the services of a competent person should be sought. FROM A DECLARATION OF PARTICIPANTS JOINTLY ADOPTED BY A COMMITTEE OF THE AMERICAN BAR ASSOCIATION AND A COMMITTEE OF PUBLISHERS.

LIBRARY OF CONGRESS CATALOGING-IN-PUBLICATION DATA
Dams : impacts, stability and design / edited by Walter P. Hayes and Michael C. Barnes.
      p. cm.
   Includes index.
   ISBN 978-1-60692-618-5 (hardcover)
 1. Dams. I. Hayes, Walter P. II. Barnes, Michael C.
   TC540.D3726   2009
   627'.8--dc22
                                                        2008055488

Published by Nova Science Publishers, Inc. □ ✛ □ □ *New York*

CONTENTS

PREFACE

This book describes the impact, stability and designs of dams, which are based upon structural, geotechnical and hydraulic properties. The impact of the dam structure on the upstream and downstream catchments are considered: e.g., sediment trap, fish migration, downstream water quality, and modifications of the water table and associated impacts (salinity). The impact dams have on the survival of certain fish is also reviewed. For example, dams impact the survival of juvenile anadromous fishes by obstructing migration corridors, lowering water quality, delaying migrations, and entraining fish in turbine discharge. To reduce these impacts, structural and operational modifications to dams— such as voluntary spill discharge, turbine intake guidance screens, and surface flow outlets are instituted. These data on the operational and structural changes to the dams for the benefit of anadromous fish populations are examined. This book also describes the bankfull hydraulic geometry adjustments caused by check dams. The concrete-face rockfill dam (CFRD), a simple dam type which has remarkable economical, ecological and environmental benefits, is analyzed as well, as it has become one of the most widely used dam types. Finally, this book explores two aspects related to the environmental impacts that trigger the construction of hydrological correction check dams: the erosion caused within the riverbeds and the ploughing and soil movements made by the opening of paths in order to grant access to the check dams.

Chapter 1 - Dams impact the survival of juvenile anadromous fishes by obstructing migration corridors, lowering water quality, delaying migrations, and entraining fish in turbine discharge. To reduce these impacts, structural and operational modifications to dams— such as voluntary spill discharge, turbine intake guidance screens, and surface flow outlets—are instituted. Over the last six years, we have used acoustic imaging technology to evaluate the effects of these modifications on fish behavior, passage rates, entrainment zones, and fish/flow relationships at hydroelectric projects on the Columbia River. The imaging technique has evolved from studies documenting simple movement patterns to automated tracking of images to merging and analysis with concurrent hydraulic data. This chapter chronicles this evolution and shows how the information gleaned from the scientific evaluations has been applied to improve passage conditions for juvenile salmonids. We present data from Bonneville and The Dalles dams that document fish behavior and entrainment zones at sluiceway outlets, fish passage rates through a gap at a turbine intake screen, and the relationship between fish swimming effort and hydraulic conditions. Dam operators and fisheries managers have applied these data to support decisions on operational and structural changes to the dams for the benefit of anadromous fish populations in the Columbia River basin.

Chapter 2 - Strengthening of dam foundations is sometimes necessary because of potential seismic activity near the dam. A jet grouting test section was undertaken at the Mormon Island Auxiliary Dam (MIAD) near Sacramento, California during the spring and summer of 2007. Jet grouting is the high pressure injection of cement grout slurry deep into soil layers to mix material with grout in order to stabilize the soil structure. The purpose of the test section was to evaluate the feasibility of jet grouting for the foundation soils and to evaluate construction issues such as waste handling, groundwater monitoring, strength/quality of installed concrete, and biological monitoring.

Monitoring of nearby wetlands occurred because some grout materials affect water chemistry and may cause increased pH of groundwater near grout columns. Changes in local geochemistry, such as pH, may have negative effects on wetland environments and existing biota.

Monitoring was undertaken at a wetland complex near MIAD during the period of jet grouting. Results from pH measurements suggested there were no increases in pH coincident with jet grouting activities, and there were no detected impacts to aquatic invertebrates in wetlands adjacent to MIAD. Overall results suggest that jet grouting was a benign activity when water quality and nearby aquatic invertebrate assemblages were considered.

Chapter 3 - Bankfull hydraulic geometry relationships, also called regional curves, relate bankfull channel dimensions to watershed drainage area. Nevertheless, construction of structures crossing the stream, such as check dams, can alter these relationships. This paper describes the bankfull hydraulic geometry adjustments caused by check dams for ephemeral channels in South-Eastern Spain, in particular for two of its torrential catchments most modified by projects of hydrological-forest restoration: i) the Cárcavo catchment, predominantly marly and lime-marly, and ii) the Torrecilla catchment, composed of metamorphic materials. The variations of hydraulic geometry at the bankfull stage between the reaches upstream and downstream from the check dams are analyzed in relation to the characteristics and spacing of the check dams, and especially in relation to other geomorphological variables also affected. Longitudinal bed slope, bed stability indexes, textural changes of deposits, dimensions of scour holes developed downstream from the check dams, upstream sedimentary wedges and refill degree of sediments behind these cross structures have been taken into account. Different regression equations were developed for both torrential streams with the aim of better defining how check dams affect ephemeral channels. Interesting results have been obtained from the bankfull hydraulic geometry relationships versus the watershed morphometry, the individual distance between check dams and the accumulated distances by consecutive refilled dams providing a continuous flow and sediment supply.

Chapter 4 - The storage of water is essential for providing our society with drinking and irrigation water reserves. Storage along a natural stream is possible if the hydrology of the catchment is suitable. Hydrological studies provide information on the water volumes and as well as on the maximum (peak) flow in the system. Often the stream runoff does not provide enough supply all year round, and an artificial water storage system (e.g. the reservoir behind a dam) must be developed. For design engineers, it is essential to predict accurately the behaviour of hydraulic structures under the design conditions, operation conditions and emergency situations. The design approach is based on a system approach. A hydraulic structure must be analysed as part of its surroundings and the hydrology of the catchment plays an important role. Structural and hydraulic constraints interact, and the design of hydraulic structures is a complex exercise altogether. For example, the construction of a dam across a river requires a hydrological study of a stream. If the catchment can provide enough water all the year around, the risks of exceptional, emergency floods must be assessed. The design of the dam is based upon structural, geotechnical and hydraulic considerations. Political issues may further affect the site location and the final decision to build the structure. A consequent cost of the dam structure is the spillway system designed to pass safely the maximum peak flood. In addition the impact of the dam structure on the upstream and downstream catchments must be considered : e.g., sediment trap, fish migration, downstream water quality, modifications of the water table and associated impacts (e.g. salinity).

Chapter 5 - The concrete-face rockfill dam (CFRD) is a simple dam type and has remarkable economical, ecological and environmental benefits, and thus become one of the widely used dam types. As viewed from experiences on construction and operation of a series of 200 m-height scale CFRDs in China, rockfill acts as the main supporting structure and the control of embankment deformation is a critical technology. Physically, rockfill is composed of fragmental materials, such as primary discrete particles. The mass stability is developed by the friction and interaction of one particle on another rather than by any cementing agent that binds the particles together. For dense granular matter, our research team has developed a 3-D numerical code *Tsinghua DEM simulation* (THDEM) by employing rigorous contact mechanics theories. It has the capacity to examine contact force details that are normally inaccessible, and to perform rigorous parametric studies. In this work, by treating a rockfill as an assembly of discrete particles, a small-scale rockfill is constructed with 50,638 poly-dispersed particles, which takes the 233 m high Shuibuya CFRD in China as the prototype. The spatial distribution of interparticle forces and stress propagation within the rockfill are obtained. Concentrated releases of forces are found at specific locations on the slab and the foundation respectively, which indicate the possible deformation of the slab or settlement of foundation. These results would provide an insight into rheological behaviours of rockfill and deformation process of concrete slab face, from a novel viewpoint of heterogeneous distribution of forces instead of averaged stress distribution in rockfill.

Chapter 6 - The hydrological correction check dams, transversely built in the riverbeds, have as a main aim the retention of sediments arriving to the reservoir, delaying their silting and expanding their lifetime.

In this research we deal with two aspects related to the environmental impacts that trigger the construction of hydrological correction check dams: the erosion caused within the riverbeds and the ploughing and soil movements made by the opening of paths in order to grant access to the check dams.

Downstream of these check dams, an incision is produced as a result of the slope breaking provoked by the check dam. The length of the stretches cofferdamed by erosion downstream of these check dams in the riverbeds studied, fluctuates between 50 and 150 meters from the bottom of the check dam. It has been calculated that the sediments evacuated in these stretches represent between 10 and 15 per cent of the accumulated deposits in the check dams located further down, reaching in some cases 50 per cent. This erosion reduces considerably the lifetime of the check dams and question, in some cases, their own usefulness.

In the other hand, the access paths to the check dams´ construction points have an environmental cost, which in some cases, go against the aim of such constructions and are difficult to justify given the little sediment accumulation capacity of the check dam, or the low rate of erosion of the gully where it is being built. The study carried out in the basin of River Segura, located in the southeasterm of Spain, shows that the relation between the soils removed in the access paths and the soil retained is approximately 10 per cent, although the fluctuations are between 7 and 36 per cent. Besides, from these paths and at the heights of each check dam, some others are constructed that are directly going to the basin of the check dam, with slopes that in many cases, go over 100 per cent and finally become a source of "non natural" sediments added to the already mentioned erosion of the riverbeds. Finally, the ploughed area with bushes and pine elimination has an approximately average of 0.7 hectares per check dam, reaching 1.6 in one of the gullies.

Notice that, the erosion added effects in the riverbeds and the erosion in paths that go directly to the basin of the check dam can represent as an average, 20 per cent of the sediments accumulated in each check dam. This erosion will not be produced if check dams had not been built, and apart from this, it is necessary to get rid of almost a hectare of bushes per check dam, for building access paths, eliminating in many cases, protected species and been certain that the recovery of these bushes in a semiarid environment as the one in the river Segura basin will be produced in the long term or even it will not happen at all, the check dams construction in certain gullies is not justified.

In short, before the construction of check dams is essential to consider several aspects, because in some occasions the environmental and erosive impact produced by the construction itself can be higher to the benefits we pretend to obtain.

Chapter 7 - Southern Spain has increasing water-supply problems due to the expansion of cities, tourism, and farming areas. In 2005, the Rules Dam was constructed on the Guadalfeo River 20 kilometres from the sea in the search for improved water provisions for urban and irrigation uses. Subsequently, river discharge decreased, especially in the sector nearest the river mouth, where the river is now permanently dry. For centuries, the discharge of this river and irrigation excess (previously derived from the river) were the main sources of recharge for the Motril-Salobreña coastal aquifer, formed by the deposition of sediments at the mouth of the Guadalfeo River. The interruption to river flow is likely to decrease these inputs to the aquifer, most probably causing a reduction in aquifer resources that would then affect the behaviour of the hydrogeological system. Therefore, the study of the influence of these factors on the water table could provide crucial information on the dam impacts on groundwater. One of the main secondary effects due to a water-table decrease in coastal aquifers is saltwater intrusion. Along the neighboring coastal aquifers, seawater encroachment has been detected, but in the Motril-Salobreña aquifer, the influence of the river maintained (until recently) optimal quality and quantity of groundwater. The study of marine intrusion is very important due to possible changes; the main problem is the great aquifer thickness (more than 200 metres) and the lack of boreholes for the measurement of groundwater conductivity. The combination of two geophysical techniques has been very useful for the quantification of the state of this aquifer. The monitoring of the possible effects from the decrease in the Guadalfeo River flow, the water-table drop and, finally, saltwater encroachment, as the main hydrogeological consequences of the Rules Dam, may allow the dam's negative impacts to be minimised by planning for sustainable management.

Chapter 8 - The application of the ecological concept of natural flow regimes to the study of hydrologic impacts of dams necessitated simultaneous analysis of numerous hydrologic variables and identification of the factors influencing the extent of hydrological changes of these variables downstream from dams. Due to some of their weaknesses, the simple and multiple regression methods did not make the most of this concept's contribution to the study of hydrologic impacts of dams. To mitigate these weaknesses, we apply canonical correlation analysis. We correlated 7 hydrologic variables which define the fundamental characteristics of annual maximum and minimum streamflows with 8 explanatory factors for 62 stations for which streamflows are measured downstream from dams in Quebec. This analysis identified the factors influencing the magnitude of the hydrologic changes in the characteristics of the annual maximum and minimum streamflows downstream from dams in Quebec.

- Streamflow magnitude and frequency are mainly influenced by the watershed size. The changes in annual maximum streamflow are greater for large watersheds (> 10,000 km²) than for small watersheds concerning annual minimum streamflows. This factor may be sufficient to estimate the streamflows downstream from dams deprived of hydrologic data, except for the lowest minimum streamflows.
- The timing of annual maximum streamflows is mainly influenced by the type of regulated hydrologic regime (dam management mode) while the timing of annual minimum streamflows is mainly influenced by the degree of water storage.
- The timing variability of annual minimum streamflows is influenced by the cumulative maximum capacity of the reservoirs in the system (number of dams built on the same watercourse).

The variability and asymmetry of the magnitude are not influenced by any explanatory factor analyzed.

Chapter 9 - Due to ever-increasing population, water scarcity, and pressing issues of environmental sustainability, Asian rice farmers are under considerable pressure for sustainable increase in rice production by using less water. It is widely believed that an increase in the water use efficiency through integrated crop management holds promise of increased yields and water productivity. The so-called System of Rice Intensification (SRI) is attracting favorable attention of farmers and governments in Asia and elsewhere. It is assumed that a healthier and larger root system can be induced by using "SRI principles" in a water-limiting environment giving positive impacts on grain yield. The cultural practices that characterize SRI includes rapid and shallow transplanting of younger seedlings, at wider spacing and maintaining alternate wet and dry condition or preferably just moist conditions during the vegetative stage. This chapter reviews the biological mechanisms of water-saving agriculture and its relation to SRI cultural practices. It presents some research findings on the rice plant's adaptive trait which could be utilized to manage crops under limited water application. In addition, the details of on-farm studies carried out in some of the rice growing countries of Southeast Asia using a participatory action research approach are included in this chapter, which asserts the need for integrating science, people and policy makers for better and sustainable water management in Asia and elsewhere.

Chapter 10 - The fluid memory is the most important yet most neglected feature in considering fluid flow models, since it represents the history of the fluid and how it will behave in the future. This paper introduces a stress-strain model where all the probable properties have been incorporated with viscous stresses. The derived mathematical model introduces the effect of temperature, the surface tension, pressure variations and the influence of fluid memory on the stress-strain relationship. The part of the stress-strain formulation related to the memory is taken into account and we obtain the variation of it with time and distance for different values of α. The notation α is shown the effect of memory and varied $0 \leq \alpha < 1$. The zero value shows no memory effect while the unity values of α shows the most extreme case of the effect of memory. The fluid memory effects as a function of space and time are obtained for the fluid in a sample oil reservoir. The dependency of fluid memory is considered to identify its influence on time. As pressure is also a function of space, the memory effects are shown in space with pressure gradient change. The computation indicates that the effect of memory cause a nonlinear and chaotic behavior for stress-strain relation. This model can be used in reservoir simulation and rheological study, well test analysis, and surfactant and foam selection for enhanced oil recovery.

In: Dams: Impacts, Stability and Design
Editors: Walter P. Hayes and Michael C. Barnes

ISBN 978-1-60692-618-5
© 2009 Nova Science Publishers, Inc.

Chapter 1

Reducing the Impacts of Hydroelectric Dams on Juvenile Anadromous Fishes: Bioengineering Evaluations Using Acoustic Imaging in the Columbia River, USA

Gary E. Johnson, Gene R. Ploskey,
John B. Hedgepeth, Fenton Khan, Robert P. Mueller,
William T. Nagy, Marshall C. Richmond and Mark A. Weiland
Pacific Northwest National Laboratory, Richland, WA, USA

ABSTRACT

Dams impact the survival of juvenile anadromous fishes by obstructing migration corridors, lowering water quality, delaying migrations, and entraining fish in turbine discharge. To reduce these impacts, structural and operational modifications to dams—such as voluntary spill discharge, turbine intake guidance screens, and surface flow outlets—are instituted. Over the last six years, we have used acoustic imaging technology to evaluate the effects of these modifications on fish behavior, passage rates, entrainment zones, and fish/flow relationships at hydroelectric projects on the Columbia River. The imaging technique has evolved from studies documenting simple movement patterns to automated tracking of images to merging and analysis with concurrent hydraulic data. This chapter chronicles this evolution and shows how the information gleaned from the scientific evaluations has been applied to improve passage conditions for juvenile salmonids. We present data from Bonneville and The Dalles dams that document fish behavior and entrainment zones at sluiceway outlets, fish passage rates through a gap at a turbine intake screen, and the relationship between fish swimming effort and hydraulic conditions. Dam operators and fisheries managers have applied these data to support decisions on operational and structural changes to the dams for the benefit of anadromous fish populations in the Columbia River basin.

INTRODUCTION

Hydroelectric development can have serious impacts on fish populations (National Research Council 1996). Fish movements can be delayed by reduced velocities in reservoirs relative to a free-flowing river (Berggren and Filardo 1993; Raymond 1968; Raymond 1979). Injuries or death can result from turbine blade strike or abrupt pressure changes, among other mechanisms (Mathur et al. 1996). After passing through a dam, fish can be disoriented, increasing their vulnerability to piscivorous and avian predators (Coutant and Whitney 2000). All of these impacts are especially a problem for migratory species. For anadromous fishes, such as the family Salmonidae, the juvenile life stage, during which fish move from freshwater to saltwater, is impacted by dams. The adult stage of catadromous fishes, such as eels (genus *Anguilla*) that migrate downstream to oceanic spawning grounds, also can be impacted by dams (Haro et al. 2000). Dams also can impede upstream migrations of fishes (Keefer et al. 2004). Structural and operational engineering solutions to protect fish have been pursued and evaluated through research. Acoustic imaging is an evolving and versatile research technique that is providing information to decision-makers who are responsible for reducing the adverse impacts of Columbia River dams.

Background

In the Pacific Northwest (Figure 1), the impacts of the Federal Columbia River Power System (FCRPS) on fishes are wide ranging. Juvenile salmonids (*Oncorhynchus kisutch, O. mykiss, O. nerka ,*and *O. tshawytscha*) can be injured or killed during dam passage (National Research Council 1996). Lamprey (*Lampetra tridentata*) can be impinged on turbine intake screens (Moursund et al. 2001). Sturgeon (*Acipener spp.*) migrations can be impeded or blocked (Beamesderfer and Nigro 1992). Bull trout (*Salvelinus confluentus*) habitat can be cut off by dams (Neraas and Spruell 2001). Reservoirs behind the dams can provide habitat conditions favorable to the northern pikeminnow (*Ptychocheilus oregonensis*) and American shad (*Alosa sapidissima*), which eat and compete with juvenile salmon, respectively (Peterson 2001). During summer months, water temperatures in rivers and reservoirs generally rise as solar energy input increases (Wetzel 2001), which can then affect fish predation rates (Petersen and Kitchell 2001). Furthermore, FCRPS flow regulation, along with diking and shoreline management between the last dam and the ocean, have resulted in a loss of connectivity to shallow water rearing habitats for juvenile salmonids in the tidal fresh and estuarine waters of the Columbia River (Bottom et al. 2005). Therefore, much research has been undertaken to support efforts to improve the survival of fish populations impacted by FCRPS dams.

The U.S. Endangered Species Act (ESA) is a major driver for this research because 13 salmonid stocks are currently listed as threatened or endangered in the Columbia River basin (National Oceanic and Atmospheric Administration [NOAA] 2008). There are over 60 dams in five Pacific Northwest states that block over 4,600 stream-miles of historical habitat (Northwest Power and Conservation Council 2000) and affect fish using the remaining habitat to reproduce, rear, and migrate. Fourteen of the 60 dams are FCRPS projects. Because the FCRPS impacts ESA-listed fish species, NOAA Fisheries and the Action Agencies

(Bonneville Power Administration [BPA], U.S. Bureau of Reclamation, and U.S. Army Corps of Engineers [USACE]) undergo a consultation process that results in a Biological Opinion (BiOp) on FCRPS operations. Such BiOps were issued in 1994, 1995, 2000, and 2004; NOAA (2008) is the current BiOp. The BiOps have included mandates for actions at the mainstem Columbia and Snake river dams, such as turbine intake occlusions, intake screens, and surface flow outlets, to increase the survival rates of juvenile and adult salmon migrating through the FCRPS. Depending on the action and the requisite objectives for evaluation, various research techniques are used, including acoustic imaging for direct observations of fish.

Acoustic imaging bridges the gap between fisheries hydroacoustics and underwater video systems. Scientific-quality fisheries hydroacoustic systems can detect acoustic targets at long ranges, but cannot record the shapes of targets as well as video systems (Mueller et al. 2006). Optical video systems can image fish in clear water, but provide limited image quality at low-light or high-turbidity levels. Underwater acoustic imaging cameras use acoustic lenses to form very narrow beams (e.g., 0.4° by 10.8°) that provide high resolution of targets. In fact, lateral images of fish within about 15 m of the camera are clear enough to reveal fish undulating as they swim. The camera also allows users to measure fish length and discriminate the fish's head from its tail. Acoustic cameras are advantageous because they have a relatively large sample volume (e.g., 29° W by 11° H) compared to fisheries hydroacoustics (6° to 15° circular) and they allow users to discriminate individuals within fish schools. Large raw data file sizes and poor species recognition are disadvantages of these cameras.

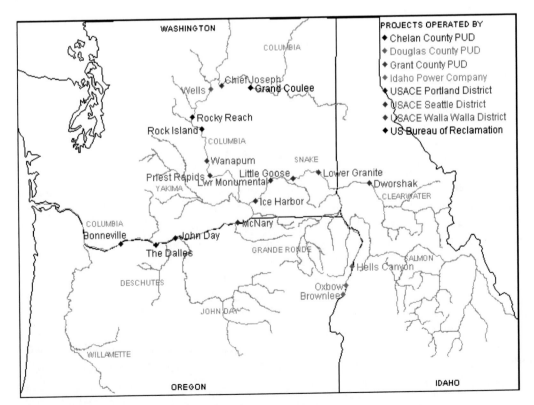

Figure 1. Columbia and Snake River Dams in the United States.

Acoustic imaging technology was developed originally for military applications, such as detecting mines, but it soon found use in fisheries science. For example, applications include enumerating adult salmon migrating to spawning areas in Alaskan rivers (Maxwell and Gove 2004; 2007) and the Fraser River (Holmes et al. 2006); imaging and counting salmon redds below Bonneville Dam on the Columbia River (Tiffan et al. 2004); river bottom profiling to improve escapement estimates (Maxwell and Smith 2007); fish behavior monitoring at tide gates in Australia (Pease and Green 2007); fish imaging under ice in the Arctic (Mueller et al. 2006); and assessing fish behavior near baited fishing trawls (Rose et al. 2005). Boswell et al. (2008) described approaches for automated tracking of acoustic image data. The first published application of acoustic imaging for fisheries research at dams was by Moursund et al. (2003).

Objective

The objective of this chapter is to describe the background, objectives, methods, results, and management implications of research concerning four major topics addressing impacts on juvenile salmonids at the Bonneville and The Dalles dams on the Columbia River. The research topics are 1) fish behaviors — observing fish movements, orientations, densities, etc. to visualize how the fish, including predators such as northern pikeminnow, are behaving in areas of interest near the dams; 2) fish passage rates — estimating passage rates of juvenile salmonids into a portal to know how many fish are lost at that portal and if the structural improvements to prevent that loss are working; 3) fish entrainment zones — estimating the probability that fish will be entrained into a portal to quantify the zone of influences of the portal, such as a surface flow outlet; and, 4) fish and flow relationships — determining quantitative relationships between fish swimming effort, as derived from fish movement and water velocity data, and hydrodynamic conditions, including velocity, acceleration, turbulence, and strain, to establish engineering design guidelines for fish passage structures.

STUDY AREA

To meet the objective of this chapter, we report acoustic imaging research conducted during the last six years on juvenile salmonids migrating downstream through the Bonneville and The Dalles dams (Table 1). The research demonstrates the breadth of the problem of dam passage and some of the solutions designed to reduce impacts on the passing salmonids.

The Columbia River, with a drainage basin area of 660,480 km^2 (Simenstad et al. 1990), has the fourth highest average discharge and the sixth largest watershed in the United States, based on U.S. Geological Survey (USGS) analysis that included the Great Lakes/St. Lawrence, Yukon, Mississippi, Missouri, and Ohio rivers (USGS 1990). Historically, estimates of unregulated Columbia River flow range from a minimum of 2,237 m^3/s in the fall to maximum flood flows of over 28,317 m^3/s during spring freshets (Sherwood et al. 1990). Since the 1930s, however, the timing of the Columbia River's discharge has been progressively regulated because of construction and operation of hydroelectric dams. These dams reduce spring freshet flows and increase fall and winter flows in the river's main stem and tributaries to meet demands for power generation, flood control, and irrigation.

The FCRPS dams are owned and operated by the USACE and the U.S. Bureau of Reclamation, while the BPA owns the FCRPS transmission system and markets and distributes the electrical energy the system produces (>25,000 MW). Construction and operating costs for the FCRPS are paid by BPA, using collections from power and transmission sales. Fisheries research at USACE-operated dams is funded by Congressional appropriation. The FCRPS is the largest producer of hydroelectric power in the United States.

Two hydroelectric projects in the FCRPS—the Bonneville and The Dalles dams (Table 2)—are the subject of this chapter. Operated by the USACE Portland District, both dams are run-of-river projects that generally draft to lower reservoir limits during the day to meet power demand and refill to upper limits during the night when possible. During spring and summer, 25 to 40% of daily average discharge at these dams is voluntarily discharged (as spill) through the respective spillways to protect juvenile salmonids migrating downstream from passing into turbines.

Table 1. Study Topics, Years, Locations, and Subjects

Topic	2002	2003	2004	2005	2007
Behaviors	TDA turbine intake occlusions	--	--	--	--
Passage Rates	--	B2 screen gap-closure device	B2 screen gap-closure device	--	--
Entrainment Zones	--	--	B2 corner collector SFO	B1, B2, TDA sluiceway SFOs	--
Fish/Flow Relationships	--	--	--	--	TDA sluiceway SFO

B1 = Bonneville First Powerhouse; B2 = Bonneville Second Powerhouse; SFO = surface flow outlet; TDA = The Dalles Dam.

Table 2. Project Data

Feature	Bonneville Dam	The Dalles Dam
Distance from River Mouth	247 km	309 km
Completion Date and Number of Turbine Units	1938 (B1: 10 turbine units); 1982 (B2: 8 turbine units)	1960 (14 turbine units); 1973 (8 additional turbine units)
Generating Capacity	1,050 megawatts	1,780 megawatts
Powerhouse Hydraulic Capacity	8,155 m^3/s	10,619 m^3/s
Juvenile Fish Passage Facilities	B1 sluiceway; B2 intake screens on all turbine units and a sluiceway "corner collector"	Sluiceway
Adult Fish Passage Facilities	Fish ladders with fish-counting stations at both powerhouses.	Fish ladders with fish-counting stations

Obtained on May 21, 2008 from www.nwcouncil.org and www.bpa.gov.

Obtained from Google Earth.

Figure 2. Satellite Image of Bonneville Dam. Flow is from right to left.

Bonneville Dam

Of the hydropower dams in the FCRPS, Bonneville Dam is closest to the ocean. As such, it passes more downstream-migrating juvenile salmonids that survive to the ocean than any other FCRPS dam. Bonneville Dam is comprised of two powerhouses, a spillway, and a navigation lock (Figure 2), all separated by islands in the river. The thalweg is in the spillway and Bonneville First Powerhouse (B1) approach channels. B1 has 10 turbine units, each with 3 intakes. The spillway has 18 bays with flat vertical gates and is located in the central portion of the Bonneville complex. The Bonneville Second Powerhouse (B2) consists of 8 turbine units, each with three intakes, and is the priority source for power generation when choosing between the two powerhouses because of its state-of-the-art juvenile fish passage facilities.

A concerted effort has been made to improve passage conditions for downstream migrants at the project, because of its downstream position in the system and because it has had consistently low estimates of fish passage efficiency (Gessel et al. 1991). B1 has an ice and trash sluiceway that is opened (42 m³/s) to pass juvenile salmonids. Turbine intake screens to divert juvenile fish into a bypass system were removed in 2004 because of low survival rates associated with the screen bypass system (Ploskey et al. 2006). All turbine intakes at B2 have modified gatewell slots with submerged traveling screens to divert fish into the low-flow bypass. Gessel et al. (1991) discuss the history of the development of intake screens at B2. Modifications include a turning vane to guide more water up into the gatewell slot, expanded vertical barrier screens, and gap-closure devices that reduce the gap between the top of the screen and intake ceiling. The outfall for this bypass was extended 3.2 km downstream to deposit fish in fast-moving water away from known predator habitats. In addition, the sluiceway at B2 was modified to pass 144 m³/s and convey fish 1.6 km downstream to a high-flow outfall. For more information about Bonneville Dam, see publications by Sweeney et al. (2007) who integrated technical literature on the development and evaluation of surface flow outlets and Ploskey et al. (2006) who summarized fish passage and survival studies.

The Dalles Dam

The Dalles Dam, located at river kilometer 309, is the second closest dam to the Pacific Ocean in the FCRPS. The Dalles Dam includes a navigation lock, a spillway perpendicular to the main river channel, and a powerhouse parallel to the main river channel with non-overflow dams on each side (Figure 3). The Dalles Dam is the only Portland District project that has the powerhouse running parallel instead of perpendicular to the main channel of the Columbia River. Full pool elevation is rated at 49 m above mean sea level (msl) and minimum operating pool elevation is 47 m msl. The thalweg intersects the dam at the eastern end of the powerhouse and, although there are deep areas immediately in front of the powerhouse, much of the forebay is relatively shallow (<20 m deep). The powerhouse is 637 m long with 22 main turbine units, each with three intakes. Flow through the main units can range from about 255 to 396 m³/s depending upon efficiency, head, desired power output, and other factors. The 421-m-long spillway is composed of 23 bays with 15-m-wide radial gates numbered sequentially from the Washington side to the Oregon side. The spillway was modified during winter 2003/2004 to include a spillwall 59 m long that divides the downstream stilling basin between Bays 6 and 7.

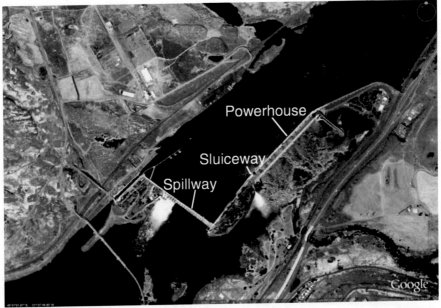

Obtained from Google Earth.

Figure 3. Satellite Image of The Dalles Dam. Flow is from right to left.

The primary structure to pass juvenile salmonids at the powerhouse is the ice and trash sluiceway, a channel that extends the entire length of the forebay side of the powerhouse. The sluiceway has three 6-m-wide entrance gates positioned over each of the 22 turbine units. Water enters the sluiceway channel from the forebay when gates are moved off the sill at elevation 46 m. A maximum of six sluiceway gates can be opened at any time before reaching the hydraulic capacity of the channel (~127 m^3/s). Flow into the sluiceway is dependent on forebay elevation and the number and location of open gates. Overall, sluiceway discharge is a relatively small proportion of total project discharge (~2%). The ice and trash sluiceway has long been operated to pass juvenile salmonids at The Dalles Dam. For additional information, Ploskey et al. (2001) and Johnson et al. (2007) have described fish passage research at The Dalles Dam from 1998–2001 and 2002–2005, respectively.

River Discharge and Fish Migration Characteristics

Columbia River discharge typically peaks in the May–June time frame and is lowest in September (Figure 4). Annual discharge is variable; for example, daily average discharge measured at The Dalles Dam during 2001 was 3,058 m^3/s (67% of the 10-year, 1998–2007, average), whereas discharge during 2006 was 5,097 m^3/s (111% of the 10-year average). The hydrographs for Bonneville and The Dalles dams are similar because they are run-of-river projects on a 62-km stretch of the Columbia River.

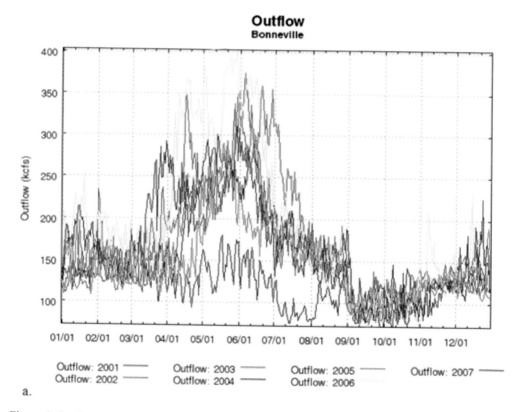

Figure 4. Continues on next page.

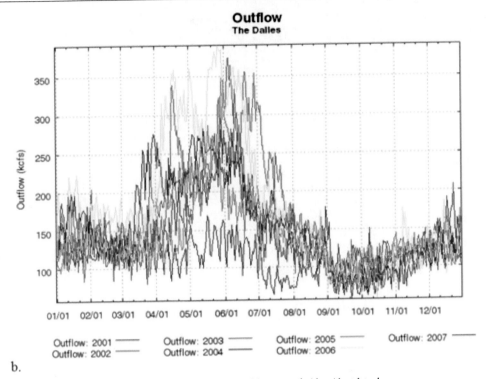

b.

Obtained on May 21, 2008 from http://www.cbr.washington.edu/dart/dart.html.

Figure 4. River Discharge from 2001 through 2007 for Bonneville (a) and The Dalles (b) Dams.

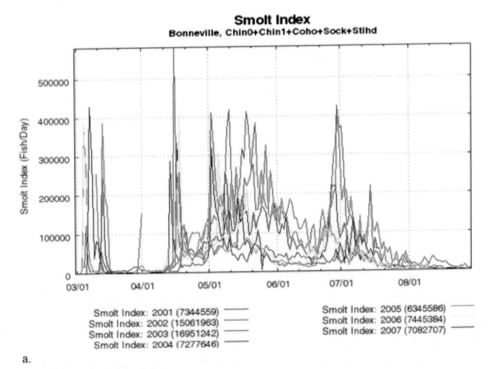

a.

Figure 5. Continues on next page.

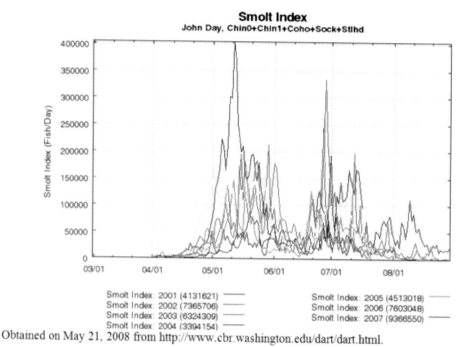

Obtained on May 21, 2008 from http://www.cbr.washington.edu/dart/dart.html.

b.

Obtained on May 21, 2008 from http://www.cbr.washington.edu/dart/dart.html.

Figure 5. Species Composition and Run Timing from 2001 through 2007 for Bonneville (a) and John Day (b) Dams. Data for John Day Dam represent The Dalles Dam because smolt monitoring is not conducted at The Dalles Dam.

Downstream migrant salmonids in the study area include yearling (stream-type) Chinook (*Oncorhyncus tshawytscha*), coho (*O. kisutch*), and sockeye (*O. nerka*) salmon, as well as yearling steelhead (*O. mykiss*) and subyearling (ocean-type) Chinook salmon. These fish originate from both wild and hatchery production. Yearling migrants have reared for over a year in their natal streams before migrating to the ocean in the spring. Subyearling salmon migrate downstream during summer after spending less than a year in natal waters. Peaks in passage during spring and summer reflect these life history patterns (Figure 5). Run-timing curves differ among the two subject dams primarily because of hatchery releases in the reservoirs immediately upstream.

GENERAL ACOUSTIC IMAGING METHODS

To study fish behaviors, passage rates, entrainment zones, and fish and flow relationships, we used an acoustic imaging device called the Dual-Frequency Identification Sonar (DIDSON) that was developed by the Applied Physics Laboratory at the University of Washington for the Space and Naval Warfare Systems Center harbor surveillance program (Belcher and Lynn 2000). The DIDSON (see Table 3 for specifications) uses acoustic lenses to form narrow beams (0.3° to 0.4°) when transmitting sound and receiving echoes (Belcher et al. 1999). Near-photographic-image clarity of individual objects is possible because the field of view is composed of multiple narrow beams. Even in turbid water, the instrument produces a near-field (within 20 m) image resembling that of a video camera. To be clear, the high resolution is only in the two dimensions of the plane in which the instrument is aimed. Unlike single- and split-beam hydroacoustic transducers, however, an acoustic camera can be aimed obliquely to a flat surface and still record fish swimming very near that surface, e.g., within a 2-m layer of sluiceway flow (Figure 6), because of the multiple beams. The DIDSON's capability to accurately image was validated for adult salmon (>50 cm) by Holmes et al. (2006), Maxwell and Gove (2007), and for juvenile salmon (>9 cm) by Moursund et al. (2003). The DIDSON was a useful assessment tool for our study because it is highly portable, requires minimal site preparation to deploy repeatedly, and does not rely on handling and tagging fish, which might alter their behavior. The DIDSON is one of the few tools that can be used to record behavior in near real time (7 to 10 frames per second). In contrast, the frequency of position estimates based on tracking of acoustically tagged fish is much lower (once every 3 to 10 seconds) at typical tag transmission rates used in survival studies. With the DIDSON, we could distinguish juvenile salmonids from other targets, observe their behavior unobtrusively, and track their movements.

The equipment to collect acoustic imaging data typically consisted of the DIDSON acoustic camera, single- or dual-axis rotators, mounts, cables, and computers. The DIDSON was usually mounted to a custom aluminum trolley that was coupled to a steel I-beam attached to the dam (Figure 7). The trolley was moved manually or electrically using a winch and davit. The underwater single- or dual-axis rotator allowed us to aim the DIDSON with high accuracy (± 1° azimuth) on the pan and tilt. The aiming angles (in degrees) from the rotator were written into the DIDSON data stream frame by frame using a serial data-acquisition module. The studies in this chapter used both high and low frequency modes, depending on objectives. The ping rate was six to seven frames per second. Separate data files generated sequentially at 10-minute intervals were saved to an external hard drive. The data recorded in the field were played back in the laboratory for processing. Processing involved visual recognition of fish targets. We took systematic notes on time, size, range, behavior, orientation, and movement direction. The background and objectives of the four major research topics addressing impacts on juvenile salmonids; methods for filtering, manual tracking, and automated tracking; results; and associated management implications are presented in the following sections.

Table 3. Specifications for the Standard DIDSON

Parameter	Low Frequency Mode	High Frequency Mode
Frequency	1.0 MHz	1.8 MHz
Number of Beams	48	96
Individual Beamwidth (2-way)	0.4° H x 11° V	0.3° H x 11° V
Spacing Between Beams	0.6°	0.3°
Composite Beamwidth	29° H x 11° V	same
Maximum Range	40 m	12 m
Average Source Level	205 dB re 1 uPa @ 1 m	same
Frame Rate	4-21 frames/s	same
Dimensions	30.7 x 20.6 x 17.1 cm	same
Power Consumption	30 Watts	same

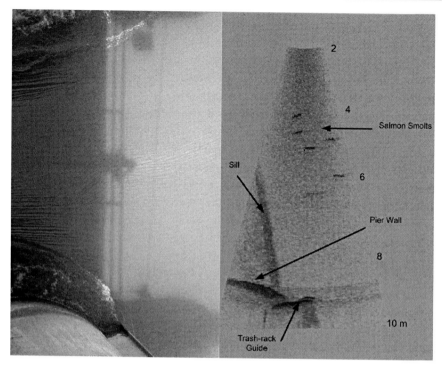

Figure 6. Aerial (left) and DIDSON (right) Views of a Sluiceway Entrance at The Dalles Dam. Details in the acoustic image include the trashrack guide slot and trashrack, pier nose, occlusion guide slot, and smolts.

Figure 7. DIDSON Mounted to a Pan-and-Tilt Rotator and Trolley Attached to a Steel I-Beam.

FISH BEHAVIORS

Fish behaviors are the movements fish make in response to environmental and innate stimuli. Generally, the research intent is to observe fish, noting length, swimming movement, orientation to flow, and schooling action, and, for our studies, differentiating juvenile salmonids from predators. Resource managers use fish behavior data to modify structures and operations at dams to make them safer for fish. Research at dams, though, necessitates dangerous deployments in fast-moving water from high perches between the dam's deck and the underwater sampling location. Besides development of the scientific-quality acoustic imaging device itself, the major advance for application of acoustic imaging at a hydroelectric dam for the purpose of examining fish behavior was the integration of the acoustic camera with a pan-and-tilt rotator and deployment apparatus. This allowed us to collect data underwater from the forebay face of a dam. Fish behavior research was presented for The Dalles Dam powerhouse during 2002 (Johnson et al. 2003).

Background and Objective

During 2002 at The Dalles Dam, prototype turbine intake occlusion plates with J-shaped extensions were evaluated as a new means of preserving juvenile salmon. The occlusion plates covered the upper half of the intakes at Main Units (MUs) 1 through 5 of the 22 turbine units at the dam. When coupled with J-extensions protruding 8 m from the bottom of each plate, the "J-occlusions" were intended to cause the turbines to draw water from deeper in the forebay than would otherwise be the case. The bioengineering premise was that deepening the turbine flow net would decrease the entrainment into turbines of juvenile migrants that naturally distributed vertically in the upper part of the water column, thereby increasing smolt survival. However, resource managers were concerned that the presence of the J-occlusions on the face of the dam might provide habitat for predators of juvenile salmonids or otherwise affect smolt behavior. Thus, one of the objectives of the J-occlusion study at The Dalles Dam during 2002 was to assess the presence and/or absence and behavior of juvenile salmonids and predator fishes in the vicinity of the J-occlusions during IN and OUT treatments. A system of hoists was installed so that each occlusion could be raised and lowered.

Methods

In this 2002 study, we used the acoustic imaging technique to quantify the behavior and number of juvenile salmonids and predators with emphasis on interactions between predator and prey. The DIDSON was attached to a steel I-beam that was welded to the J- occlusion guide frame at the main pier nose of MUs 1-2 and 3-4. The section of I-beam on MU 1-2 extended down to the floor of the J- occlusions (El. 32 m above msl) and the section on MU 3-4 extended down to an elevation of 26 m above msl. The section of I-beam on MU 1-2 extended down to the floor of the J- occlusions (El. 32 m above msl) and the section on MU 3-4 extended down to an elevation of 26 m above msl. Figures 8 and 9 show forebay and plan views of the DIDSON's coverage zone.

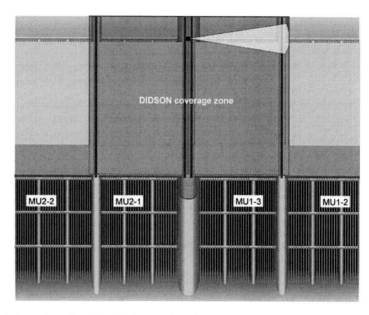

Figure 8. Front View of the Sampling Volume. The width of the DIDSON coverage zone was about 40 m.

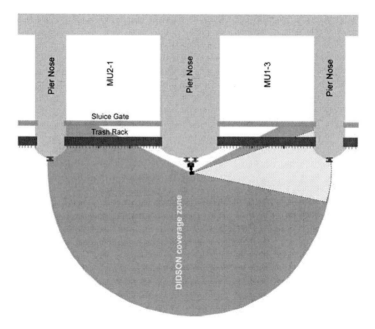

Figure 9. Plan View of Sampling Volume. The diameter of the DIDSON coverage zone was about 40 m.

Sampling for the spring and summer study was allocated between the two deployment locations at MUs 1-2 and 3-4. The sampling dates from April 26 to May 15 and from June 19 to July 12 were assigned to this project prior to the field season. A scanning regime was used to detect the presence of juvenile salmonids and predators at each location. This regime consisted of both fixed-position and active scans. This required a human operator to be present during the scans. Based on early observations, these intensive scanning periods were restricted to the diel peaks of fish occurrence from 0500 to 1000 hours and from 1700 to 2200 hours. Due to the reconciliation of the study block schedule and DIDSON availability, sampling did not occur equally between J-occlusion treatments. Rather than reducing the data set to the shortest common sampling period, rates of detection (number of fish observed per hour) were used in the analyses.

Results

Direct observations of fish as they approached the trashrack during the OUT treatment revealed a consistent behavior pattern (Figure 10), because fish did not simply approach and pass through the trashrack. Instead, fish consistently hesitated in front of the trashrack, with the trashrack functioning as a behavioral barrier. Other data showed that the water velocities were well within the swim capacity of both smolts and predators. Although this behavioral phenomenon is generally well known, it had not previously been observed *in situ* at a mainstem hydroelectric project. Some of the fish observed holding in front of the trashracks did eventually pass through the trashrack and enter the intake.

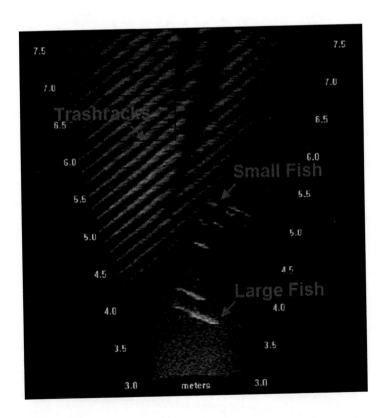

Figure 10. Large and Small Fish in Front of the Trashrack. These fish were positively rheotactic and actively avoided passage through the trashrack.

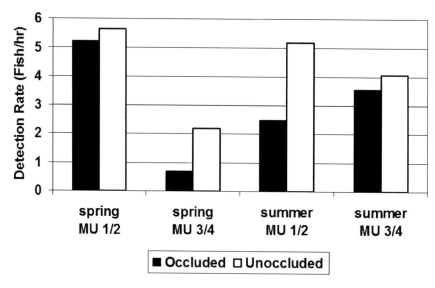

Figure 11. Number of Predator Fish Detections at Sample Location MUs 1-2 and 3-4 During the J-Occlusions IN (occluded) and OUT (unoccluded) Treatments.

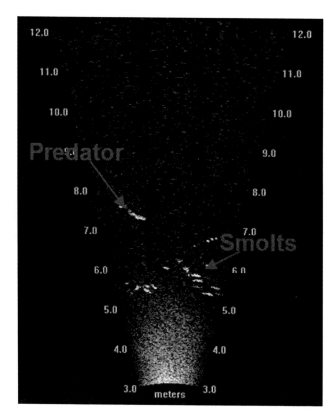

Figure 12. Simultaneous Observation of Predator and Smolts.

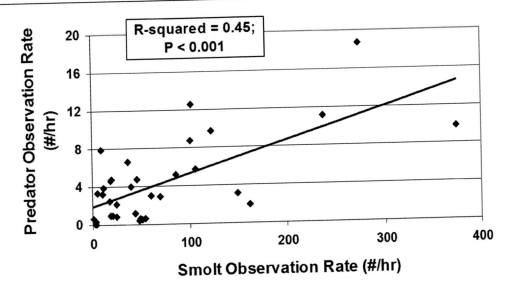

Figure 13. Relationship Between Smolt and Predator Observation Rates. Both treatments and seasons were pooled.

All of the smolts observed entering the ice and trash sluiceway passed through the center of the entrance. They appeared to avoid the piers. In addition, smolts were not committed to passage via the ice and trash sluiceway from the forebay. Smolts were free to enter and leave the area of the forebay directly in front of the open ice and trash sluiceway. Smolts, however, became entrained in sluiceway flow just over the sill at the downstream edge of the trashrack.

The number of predator fish observed per hour was similar between spring and summer study periods and higher with the J-occlusions OUT than IN (Figure 11). Predator fish at Sluice 1-3 were most likely to be observed near the sluiceway entrance staging just below the sill or near the adjacent piers. This occurred regardless of the presence of J-occlusions. A common predator staging location at both sampling locations was very near the piers. In front of MU 3-4 with the plates in and where the sluiceway gates were closed, the predators had more freedom of movement and tended to roam back and forth along the powerhouse. Only in a few instances were predator fish seen swimming near the J-occlusion floor or near the gaps between occlusion plates. In several instances predators were observed actively pursuing smolts or groups of smolts (Figure 12) and in at least one instance clearly consumed a smaller fish. Predators and prey were detected near the powerhouse at the same time. This association was true for the entire study period. We found that the predator observation rate increased as the smolt observation rate increased (Figure 13).

Management Implication

Resource managers used observations of fish behavior at the trashracks from the DIDSON study at The Dalles Dam during 2002 to support, in part, the decision to not install J-occlusions. Also, the results heightened managers' concern about predators of juvenile salmonids at the dam portals.

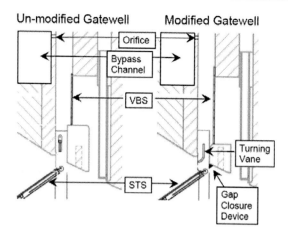

Figure 14. Cross-Section Views of Unmodified (left) and Modified (right) Gatewell Slots of B2 Intakes. VBS = vertical barrier screen; STS = submerged traveling screen. Modified features include an extended VBS, a turning vane, and gap-closure device. The gatewell is 1.3 m across.

FISH PASSAGE RATES

Fish passage rates, calculated as the number of fish passing a given location per unit of time, are used by resource managers to evaluate the performance of fish guidance structures, bypass devices, and surface flow outlets (SFOs), among other engineered fish protection measures. The key advance enabling us to estimate passage rates from acoustic image data was the development of filter algorithms to distinguish fish tracks (i.e., swim paths) from non-fish tracks. The DIDSON filter algorithms were validated against optical video data. Fish passage rates were fundamental to evaluating refinements of the turbine intake screen systems at B2 during 2003 and 2004 (Ploskey et al. 2004; 2005).

Background and Objectives

At many USACE dams on the Columbia and Snake rivers, large screens are placed in the turbine intakes to guide fish away from passing into the turbine runner by diverting them up into a gatewell where they are routed through the dam in a bypass system. A state-of-the-art intake screen bypass system for juvenile salmonids was incorporated in B2 when it was completed in 1982. However, fish guidance efficiency, calculated as the percentage of total fish passing into the turbine intakes that are guided by the screens, was less than satisfactory at 40 to 44% for yearling migrants and 18% for subyearling Chinook salmon (Gessel et al. 1991). A possible reason for low fish guidance efficiency was gap-loss, that is, fish were initially guided by screens but ultimately lost to turbine passage through the 46-cm gaps between the tops of the screens and the intake ceilings (Figure 14). To alleviate this problem, prototype gap-closure devices were installed in six intakes to direct more flow up into the gatewell slot (56% vs. 84% of total intake flow without and with the gap-closure device, respectively). The objective of research and evaluation effort during 2003 and 2004 was to estimate fish passage rates through the gap between the submersible traveling screen and the intake ceiling to determine if the gap-closure device was working as intended.

Methods

Using a DIDSON acoustic camera to sample fish passage in nine turbine intakes, we estimated fish passage rates in each intake for two adjacent routes: through the screen-ceiling gaps and up into the gatewells. We deployed a down-looking DIDSON on a laterally moving mount that traversed a horizontal beam (Figure 15), which was lowered by crane into the gatewell slots of the nine intakes at B2. The DIDSON recorded images of juvenile salmonids moving up into the gatewell and through the gap between the top of the screen and the intake ceiling. A programmable traversing mount allowed the DIDSON to sample from five lateral locations across each gatewell at 10-minute intervals. The traversing part of the 6-m-wide beam was moved by a stepper motor and controlled by custom-designed software on a laptop computer.

In spring and summer 2003, six intakes (three each at Units 15 and 17) were modified with gap-closure devices, while the three unmodified intakes at Unit 13 served as controls. In spring 2004, we sampled fish passage in the three intakes each (designated A, B, and C) of Units 13 (unmodified) and 17 (modified). Acoustic imaging was not conducted during summer 2004 because of the difficult conditions encountered sampling gap loss for subyearling fishes in summer 2003.

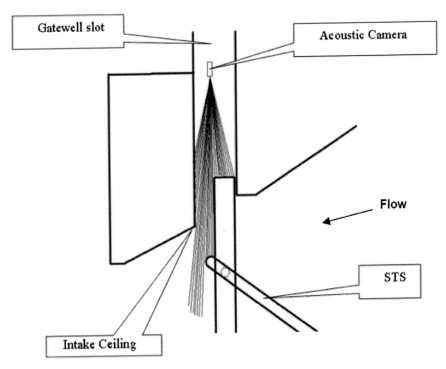

a.

Figure 15. Conintues on next page.

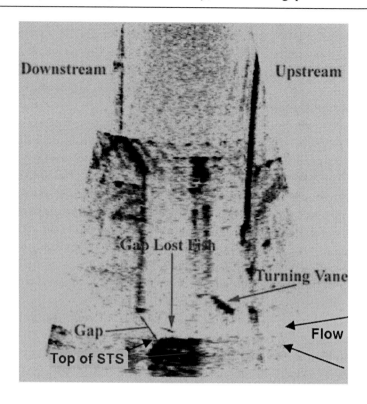

b.

Figure 15. Cross-Sectional Diagram of an Unmodified Gatewell Slot of Unit 13 at B2 Showing the DIDSON Deployment (a) and an Image from the DIDSON Deployed in Modified Gatewell 17C of B2 (b). Flow was moving from right to left below the tip of the turning vane and upward above the tip of the vane. The gatewell is 1.3 m across.

During 2003, the unmodified intakes of Unit 13 also were sampled with infrared optical cameras to evaluate the proportions of fish and non-fish objects passing through the screen-ceiling gaps. We found that fish composed just 28.6% of all objects in spring and 12.9% in summer. Experiments in a laboratory tank confirmed that the DIDSON detects echoes from the surfaces of waterlogged sticks, macrophytes, and other debris as well as from fish. We developed filters based on target size, motion, range at first appearance, and the number of frames in which a target was seen to discriminate between fish and non-fish images. Filtered data produced estimates within 6% of those obtained by multiplying unfiltered DIDSON counts by the fish fraction estimated from optical-camera data. Images from small subyearling summer migrants, however, were much more difficult to separate from those of entrained air bubbles than were images of larger spring migrants. We recommended limiting DIDSON sampling to estimate gap loss to spring when larger juvenile salmonids are present and future studies to determine the relative contribution of gap-closure devices to the gap-loss reductions.

The DIDSON data were initially processed by viewing native DIDSON files and counting the number of fish-shaped images observed moving up into the gatewell or through the gap between the top of the submersible traveling screen (STS) and the ceiling of the intake. Along with counting the number of fish-shaped images observed, we recorded the range of the object from the DIDSON, number of frames from first to last detection, number of frames in which the object was detected, shape of the object, if it was on the screen of the STS, if it undulated, and the length of the object. The data were then filtered. We deleted observations that met any of the following criteria: 1) gatewell-bound objects first detected at ranges <3 m from the DIDSON or in fewer than three frames, because fish within 3 m of the DIDSON and more than 2 m above the top of the screen had a higher probability of being re-circulated through the field of view and counted multiple times than fish detected within 2 m of the top of the screen; 2) gap-lost objects that were not undulating or crossing stream lines or that were detected in fewer than four frames in A or B intakes or in fewer than five frames in C intakes; and 3) maximum target length <70 mm and >305 mm in spring. Fish counts were expanded for each fish using the following equation:

$$EC = \frac{GW}{2\left[FR \times TAN\left(12/2\right)\right]}$$

where, EC is theexpanded count, GW is the gatewell width (6.1 m), FR is the first range of detection (m), TAN is the tangent, and 12°-long is the angle of each of the 96 0.3°-wide acoustic beams relative to the length of the GW.

The expansion increased the count of fish in the gatewell fraction relative to the gap-loss fraction because gap-loss fish were detected at a slightly greater range than were fish moving up into the gatewell.

The primary metric in this study, gap passage percentage, was derived from fish passage rates through the screen-ceiling gap and up into the gatewell. Gap passage percentage is defined as:

$$\text{Gap Passage \%} = \frac{\text{Gap Passage Rate}}{\text{Gap + Gatewell Passage Rates}} \times 100$$

Results

The 2003 study results suggested that the intake modifications at Units 15 and 17 reduced gap loss relative to rates at unmodified Unit 13 by about 67% in spring and summer. An analysis of the variance of differences in gap loss among units with modified and unmodified intakes (n = 9) indicated significantly higher gap loss in unmodified units than in modified units. In spring, the least-squares mean rate for unmodified Unit 13 was 11.4% and this rate was significantly higher (P = 0.0001) than rates of 3.8% at modified Unit 15 and 3.6% at modified Unit 17, which did not differ significantly. In summer, the least-squares mean rate for unmodified Unit 13 was 12.6% and this rate was significantly higher (P = 0.0188) than rates of 5.8% at modified Unit 15 and 3.4% at modified Unit 17.

In spring 2003, unmodified intakes at Unit 13 had higher gap losses than modified intakes at Units 15 and 17 (Figure 16). In summer, unmodified Intakes 13B and 13C also had higher gap losses than all other intakes, which did not appear to differ significantly from each other, except perhaps from the loss at Intake 17C, which was consistently low. This assessment of and among intake differences was based upon visual inspection of means and

95% confidence intervals because three nights of sampling per intake did not provide adequate statistical power for a formal test. The effect of intake modifications apparently was stronger than other potential effects such as intake location in the A, B, or C slot.

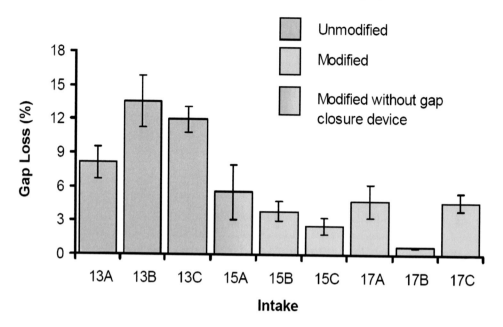

Figure 16. Gap Loss as a Percentage of Gatewell and Gap-Loss Passage at Nine Intakes Sampled Three Nights Each in Spring 2003. (Vertical bars are 95% confidence intervals about the mean.)

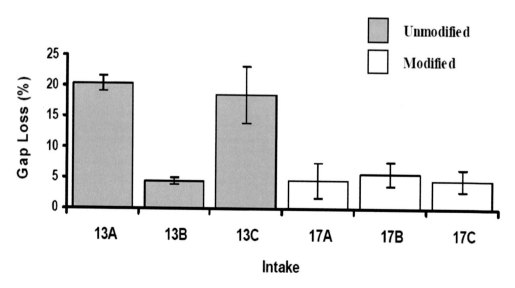

Figure 17. Gap Loss as a Percentage of Gatewell and Gap-Loss Passage at the Three Intakes of Unmodified Unit 13 and the Three Intakes of Modified Unit 17. (Each intake was sampled three nights in spring 2004. The vertical bars are 95% confidence intervals about the mean.).

During 2004, the mean gap loss of fish using filtered data (N = 3 nights) varied from 4.5 to 20% in the six intakes sampled in 2004, with the highest gap-loss estimates in the unmodified intakes. Gap loss was higher for the A and C intakes of unmodified Unit 13 than any of the three intakes of modified Unit 17 or Intake B of Unit 13 (Figure 17). The general trend was similar to what was observed in 2003 in that gap-loss estimates from DIDSON data ranged from 11.2 to 14.9% for unmodified units and from 2.9 to 5.0% for modified units (Ploskey et al. 2004; 2005). There was no correlation between gap loss and operating conditions; however, the statistical power is low because of the small sample size and the within-sample-period variability in operation levels is low and likely was insufficient to detect variation in gap loss due to operation levels.

Management Implication

Because gap passage percentages were significantly lower with the gap-closure devices in place than not, the USACE installed gap-closure devices at all 24 intakes at B2.

FISH ENTRAINMENT ZONES

The fish entrainment zone (FEZ) at a dam portal is defined as the volume of water in which fish have a 90% or greater probability of moving into the portal (Johnson et al. 2004). Entrainment zones are important because they indicate the biological extent of influence of the portal's flow field. Depending on the portal, a large FEZ could be good or bad. For example, at turbines, a large zone is bad because relatively low survival rates make turbines an undesirable passage route (Coutant and Whitney 2000). On the other hand, at SFOs, a large entrainment zone is good because the management goal is to pass fish there. The key advance enabling us to estimate entrainment zones from acoustic image data was the development of manual and automatic tracking software to process the raw data files to increase the fish sample sizes dramatically and thereby obtain databases of sufficient breadth to perform the statistical analyses that determined FEZs. FEZs were determined using acoustic imaging techniques as part of SFO development during 2004 at the B2 "corner collector" SFOs (Ploskey et al. 2005) and during 2005 at sluiceway SFOs at the B2 (Ploskey et al. 2006) and The Dalles Dam (Johnson et al. 2005).

Background and Objectives

To pass ice and trash downstream, sluiceways were incorporated into B2 and The Dalles Dam when the projects were inaugurated in 1938 and 1960, respectively. During this era (over 45 years ago), juvenile salmonid passage at mainstem Columbia dams was not a management concern. The resource management approach changed in the 1970s, as biologists realized the potential impact the hydroelectric facilities on the Columbia River could be having on salmonid populations during their downstream migration. Direct observations of juvenile salmonids being carried downstream over the sluiceway gates when they were opened to pass ice and trash led to studies of this route, a type of SFO (Johnson and Dauble 2006), as a means to divert juvenile fish from the turbines (Michimoto 1971). The first official management action to protect smolts at mainstem Columbia River dams was opening the sluiceways at B1 and The Dalles Dam during annual downstream migrations starting in the late 1970s. In fact, juvenile salmonid passage at sluiceways was the genesis of the current $100 million investment in SFOs at mainstem USACE dams on the Columbia and Snake rivers. To manage risk associated with this investment, managers require detailed information about fish movement patterns, such as those portrayed by FEZs at sluiceway entrances.

Previous studies at The Dalles Dam used fish movement data from an active fish-tracking sonar (Hedgepeth et al. 2002a) to determine entrainment zones for the sluiceway flow net (Hedgepeth et al. 2002b; Johnson et al. 2004). This sonar provides three-dimensional fish tracks by aligning the axis of a split-beam transducer with a fish target. High-speed stepper motors move the transducer as the fish moves so that a tracked target remains on axis. An advantage of this approach is its relatively high angular resolution ($\pm$ 0.35°) and three-dimensional fish position data; a disadvantage is that tracking fish is difficult because of limited acoustical sampling volume determined by the split-beam phase aperture (7° X 7°). While not perfect, the acoustic-imaging technique has a broad sampling view (e.g., 11° X 29°). It can be used in the presence of turbulence because turbulence is typically ephemeral, plus fish targets and turbulence usually contrast in intensity over a broad volume. Given the advances of the acoustic-imaging technique to study fish behaviors and passage rates, we developed tracking software and applied the technique to determine FEZs.

Acoustic-imaging research at sluiceways was conducted at B2 for 18 days between March 1 and July 8, 2004, at B1 for 31 days between May 20 and July 10, 2005, and at The Dalles Dam for 29 days between May 5 and July 14, 2005. The three studies shared a common objective: determine the FEZ in the near field of sluiceway entrances for spring (yearling salmonids) and summer (subyearling salmonids) separately for day and night.

Methods

Acoustic imaging was deployed at particular locations to describe fish positions and sample their movements. Specially developed software and Markov chain analysis, respectively, were used to extract data from the raw DIDSON files and to capture fish movements to determine FEZs.

Deployment Locations and Sample Volumes

At the B2 corner collector (B2CC), fish approaching the B2CC in the south eddy at B2 were imaged with a DIDSON operating in low-frequency mode during three 24-hour periods in early, middle, and late spring and summer. The fan of 48 adjacent 0.6°-wide by 11°-deep DIDSON beams was oriented horizontally throughout the study and covered approximately 30° (Figure 18). The DIDSON was mounted on a pan-and-tilt rotator that was controlled by a technician with a joystick to keep smolts approaching the B2CC entrance or south eddy within the DIDSON's field of view as long as possible. This allowed pan-and-tilt angles to be integrated with fish in each frame of the DIDSON file. Rotator coordinates and fish positions in the sample beams were used to describe fish positions through time in three-dimensional space. Of particular interest were approach paths relative to the boundary between eddy flow and flow into the B2CC entrance.

Fish tracking was initiated randomly in two pre-selected zones where fish had a choice of entering the B2CC entrance or swimming away (Figure 18). Each zone was divided into three ranges (0-6 m, 6-12 m, and 12-18 m), and technicians were instructed to track fish from all ranges within each pie-shaped zone. Sampling upstream of the entrainment zone where fish could make a choice was important because it was there we could learn about responses to entrance conditions.

At the B1 sluiceway, we used a DIDSON to sample movements of juvenile salmonids approaching B1 Sluice 3C during spring and summer 2005. The instrument, attached to a rotator, was deployed from a barge upstream of the pier between Sluices 3B and 3C. It was aimed laterally so we could view Sluice 3C and the approach of fish upstream of the outlet. The barge included a trolley and rail system that was used to adjust the upstream/downstream position of the DIDSON at the time of installation. This was crucial for viewing the upstream edge of the sluiceway using high-resolution DIDSON capture. The barge was anchored to the pier between Sluices 3B and 3C using a Y-shaped standoff. The standoff pivoted at both ends so the DIDSON would remain at the same depth during the entire study period. The standoff also eliminated large side-to-side and rotational movements. Additionally, the barge was tethered using 1.25-cm wire rope as a safety precaution and as a secondary stabilizer. This system positioned the DIDSON about 8 m to the east of the entrance.

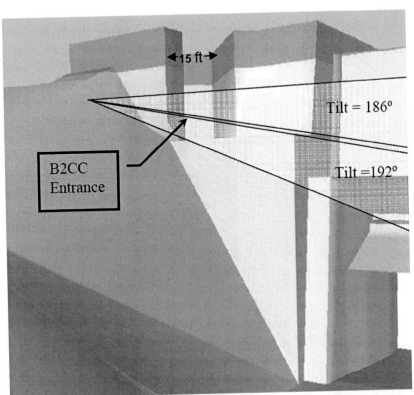

Figure 18. Front View of the B2CC Entrance Showing Two Relative Tilt Angles from the Pan-and-Tilt Rotator for Sampling Two Vertical Sections of the Water Column.

The DIDSON and rotator device were adjusted in the upstream/downstream direction by moving the trolley on the barge so that the 12-m sample range would extend just past a chain gate in the sluiceway. The chain gate sat on a sill located approximately 2 m under the water surface at normal pool elevation. We fixed the vertical axis so the DIDSON could see over the sill with the maximum beam volume and still not pick up too much surface noise produced by wind or rain. This resulted in a vertical axis return of 0° during the entire sampling period. Due to the shallow configuration of the sluice outlet, we only needed to cover the top 2–3 m of water, and this was accomplished by sampling only one vertical zone. The fan of nintey-six 0.3° beams was oriented horizontally to successively sample each of six 29°-wide, 11°-deep volumes of water immediately below the water surface for 10 minutes each (Figure 19). Sampling covered a 180° arc from the pier nose between Sluices 3B and 3C, rotating to the north, and ending almost directly upstream of Sluice 3C. Thus, the sampled area covered all flow approaching Sluice 3C.

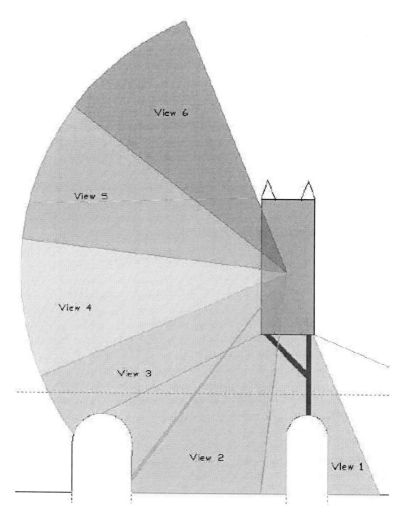

Figure 19. Plan View Showing Six 30°-Wide, 12°-Deep Sample Volumes at B1 in 2005. The six successive views were sampled sequentially for 10 minutes each. The barge with the DIDSON and rotator is represented by the rectangular object in the center of the figure.

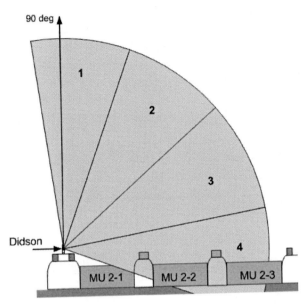

Figure 20. Plan View Showing the Four DIDSON Sampling Zones at Sluice 2 at The Dalles Dam in 2005. The light gray shading represents the approximate coverage area of the four sampling zones.

We sampled at seven frames per second using the high-frequency mode to maximize the resolution of smolt images entering the sluiceway. We intended to sample a minimum of three 3-day blocks each season. Equipment problems in spring precluded sampling before May 20th, so we were forced to acquire all three sample blocks during one 10-day period near the end the spring season. Sampling blocks were more evenly distributed in summer, and each block contained almost 5 days of sampling instead of the proposed 3-day minimum. The sampling routine was automated by a programmable motion controller. The program sampled each 29°-block for 10 minutes before rotating to the next 29° position so that a complete sample pattern took 1 hour to complete.

At The Dalles Dam sluiceway, we located a DIDSON and rotator off the pier nose between MUs 1 and 2 to sample fish approaching Sluice 2. The rotator was programmed to sample four zones near the sluiceway openings at sluiceway entrances Sluices 2-1 and 2-2. The DIDSON was deployed 2 m below the normal operating pool (El. 48 m above msl). The tilt position (vertical dimension) was set at 5° downward during the entire sampling period. Data were collected in four 29° pie-shaped horizontal zones in the surface layer along the face of the dam at Sluice 2 (Figure 20) that were sampled sequentially for 15 minutes each.

Data Processing

During 2004, a Visual Basic program was developed and used to extract spatial information from tracks of individuals and schools of fish recorded in the binary output files of the DIDSON system. The program operated to interactively identify fish tracks by boxing fish using the computer's pointing device. The relative coordinates of the opposite corners of the box were recorded in ASCII data files with the binary track file name, frame number, date, time, pan angle, tilt angle, number of fish in box, and a unique track identifier. An additional option allowing frame skipping by an amount set by the technician input number; this increased tracking production. Other options included gain and threshold settings. During manual tracking, a technician used a mouse to draw a box around fish images in successive frames to spatially and temporally define a fish track. Some fish were tracked for over 100 successive frames, so repeated mouse selection of images in frames was time consuming. In addition, technicians had to reprocess frames to track multiple individual fish in a series because only one fish per frame could be tracked at a time. When schools of fish were encountered, the entire school was tracked as a single entity and the number of fish in the school was estimated. It sometimes took at least 30 minutes to manually track a 10-minute DIDSON file.

During 2005, an autotracking software program was developed and used to extract fish tracks and rotator pan-and-tilt data from raw DIDSON files so that successive fish positions could be placed in three-dimensional real-world coordinates. In 2004, the manual tracking process was very slow and tedious, compared to the automated tracking software which processed a 10-minute DIDSON file in about 10 minutes. All fish within a DIDSON image could be tracked simultaneously, and schools were separated into individual fish. The autotracker also eliminated biases associated with having many different technicians manually process the data. The autotracking program was designed to batch process groups of DIDSON files. The program extracted fish tracks using standard image-processing algorithms and an alpha-beta tracking method (Brookner 1998). Fish track data were output as a text file containing descriptive information about each fish track, including DIDSON orientation (pan-and-tilt angles), position in the DIDSON image (range and beam), size of target (number of pixels), and amplitude of the target. A second autotracking program was developed that also output the DIDSON data file as a video showing the fish tracks as they were being processed. These videos were used to verify that the autotracking program was performing accurately and also as an aid in developing filters to remove non-smolt tracks. Filters were also developed to remove false tracks created when the rotator was panning to move the DIDSON to a different orientation and to remove a structure that was tracked as a fish. Targets with high mean amplitude or too many pixels were removed to eliminate targets that were too large to be smolts.

Prior to analysis to determine FEZs, a C-language program was used to convert the tracked fish files to fixed coordinate systems and to visualize fish tracks using Amtec Engineering's Tecplot software, while separating the data into location and day or night data sets and summary statistics. A second C-language program output a selection of the water volume's synoptic Tecplot visualization with fish tracks and for subsequent Markov chain analysis. Part of this processing included filtering tracks to accept only those longer than 3 seconds. In relative coordinates at B2 and The Dalles, the x-axis was parallel to the powerhouse and the y-axis was perpendicular to the powerhouse; the axes for the relative coordinate system at B1 are shown in Figure 21. DIDSON elevations required for adjusting the tracked fish coordinates were part of a computer program that computed both State Plane and dam-relative tracked fish coordinates. Output files were named "*.SPL" for Oregon State Plane North (OSPN) coordinates (NAD 27) but also contained relative coordinates and "*.DAT" for Tecplot software visualization in relative coordinates. Using the relative to DIDSON camera position of a fish (X, Y), its range (R), and the tilt angle θ, a single tracked fish position relative to the pointing angle of DIDSON was computed as

$$(X, Y\cos(\theta), R\sin(\theta))$$

This position was then corrected by applying rotation and translation into positions in the two coordinate systems described above. The fish tracks were then displayed and animated using Tecplot software and subjected to a Markov chain analysis.

Markov Chain Analysis

To determine FEZs, an absorbing Markov chain (Kemeny and Snell 1960) was used to capture fish movement to a particular location, the region where we considered fish were entrained into the sluiceway. A Markov chain can model continuous movement in a continuous volume when discrete time steps are chosen and volumetric cells of a sample volume are delineated over which movement probabilities for the Markov transition matrix can be calculated. The resulting Markov chain model allowed us to estimate fish movement probabilities from a given cell within the sample volume to each "absorbing" cell on the boundary of the volume.

At each site, a rectangular volume was used to apply the Markov chain analysis (Figure 21) that encompassed most of the DIDSON sample volume. The Markov sample volume was chosen to encompass a sufficient number of tracked fish to estimate movement near the sluiceway. A consideration in designing the sample volume for the Markov chain analysis was to have a sufficient number of fish reaching absorption cell boundaries. The x- and y-dimensions of cells were 0.9144 m on a side. We formed fish movement states (Kemeny and Snell 1960) for the Markov chain that corresponded to the location of each volumetric cell (Figure 21).

Markov absorbing states (Kemeny and Snell 1960), called "fates" here, were assigned on edges of the volume. By definition, movement was not possible through the surface or bottom. Fates were calculated as probabilities of absorption into cells at a particular portion of the sample volume as follows: B2CC – corner collector, southwest, reservoir and northeast; B1 Sluice 3C – sluice and non-sluice; and The Dalles Sluice 2 – sluice, east, west, and reservoir (Figure 21). If no movements to a boundary were observed, the fate was called "stagnation." Movement fates to the faces of the sample volume are simply probabilities for movements within the sample volume. In summary, the fates of interest to determine entrainment zones were the B2CC, B1 sluice, and The Dalles sluice.

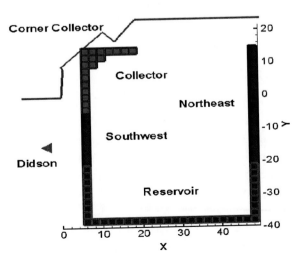

a.

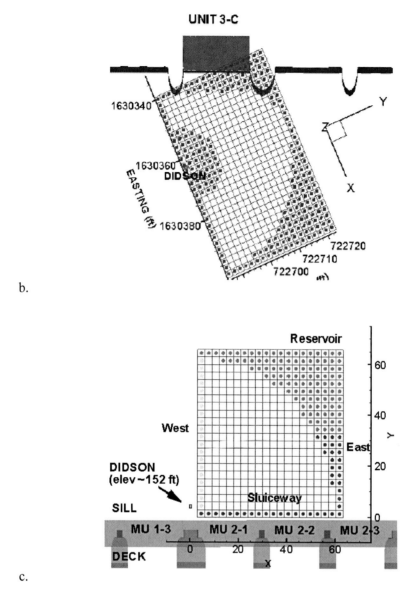

b.

c.

Figure 21. Fates Where Fish Movements Were Absorbed at Edges of the Sample Volume at B2CC (a), B1 Sluice 3C (b), and The Dalles Sluice 2 (c). (Scale: feet.)

To determine fate probabilities, we applied a Markov chain analysis (Taylor and Karlin 1998) as described for fish movement data by Johnson et al. (2004). Several assumptions were made and verified regarding the connectivity in the sample volume for the Markov model: 1) there were no absorbing non-boundary cells, that is, no interior cell's probability was equal to one; 2) exterior cells' probabilities were set to one as described above; 3) interior connectivity was not forced and relied on empirical measurements; and 4) in cells where movement data were not obtained, the closest movement was interpolated to that cell using inverse distance squared weights. Gaps between DIDSON sectors were initially filled using dithering or sector broadening. For FEZs, the data were denoted as either spring or summer based on a date division of May 30 for the end of spring and June 1 as the beginning of summer. Sunrise and sunset were used to differentiate day from night. Times were based on a table found at the website of the Astronomical Applications Department, U.S. Naval Observatory (http://aa.usno.navy.mil/data/).

A C-language program was used to construct a transition matrix and apply the Markov chain analysis. The Markov transition matrix was a square matrix the size of k x k, where k was the number of distinct cells being modeled. The ij[th] element in the ith row of the jth column of the transition matrix was the estimated probability (p_{ij}) of moving from cell i to cell j in the next time step. These probabilities were estimated by

$$\hat{p}_{ij} = \frac{n_{ij}}{n_i}$$

where, n_i = number of observations of smolts in the ith cell; and n_{ij} = number of observations where a smolt in cell i moved to cell j in the next time step.

The transition probabilities for cells (0.9 m x 0.9 m) that bordered the edges of the sample volume (e.g., sluice) were set to unity to absorb any movement that reached our defined fates. The transition matrix T was constructed using a time step of 0.5 second, using the average position (i.e., $\bar{x}, \bar{y}, \bar{z}$) during each 0.5-second interval a fish was tracked. This process required that a fish be tracked for at least 1 second before the transition matrix was amended to obtain location i from the first interval and location j from the next, and so on. The transition matrix T for one time step was used to estimate the transition probabilities for two or more time steps as T^t where t is the number of time steps, i.e., the result of multiplying the matrix T by itself t times. Matrix T^t is the transition matrix for t time steps and the transition probabilities $p_{ij}^{(t)}$ express the probability of moving from cell i to cell j in t time steps. The size of t was sufficiently large so that the tracked fish revealed an absorption state or became stagnant. The t-step transition probabilities to absorbing cells were visualized using Tecplot software by contouring the sums of each state's (each representing an x, y, z cell) probabilities for each fate.

Results

At the B2CC during 2004, over 160,500 DIDSON image frames containing fish were tracked manually. These individual frames made up 7,943 tracks (swim paths) of fish or schools of fish. Of these, 5,333 were tracked during spring data collection: 3,351 during the day and 1,982 at night. In summer, 2,610 fish were tracked: 1,409 during the day and 1,201 at night. Figure 22 shows the sums of each spatial cell's fates (states) over the corner collector and southwest absorbing edges. The largest passage probabilities were into the corner collector for both seasons. No stagnation occurred because all fish moved to the edges of the volume analyzed. Very little northeast movement along the dam face was detected. The smallest FEZ averaged 9.5 m from the dam based on passage fates of B2CC plus southwest. The FEZ varied a few feet depending upon season and time of day: spring-day was 10.4 m; spring-night was 9.8 m; summer-day was 9.1 m; and summer-night was 7.9 m.

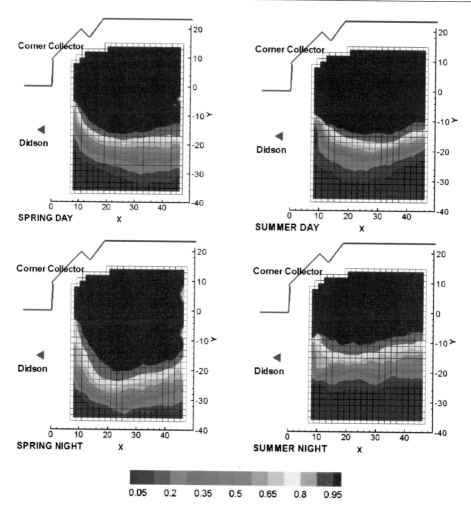

Figure 22. Contours of Fish Passage Probabilities at the B2CC for Spring and Summer 2004, Day and Night. Probabilities above are shown for the corner collector fates, and x- and y-scales are in feet.

The fate of approaching smolts, whether passing into the B2CC or moving upstream into the south eddy, could be reliably predicted from their initial location and the bulk flow in the vicinity of the B2CC. Other factors such as day, night, or time of year had little effect on approach and fate, shifting the FEZ by only about 2 m. The high effectiveness estimates for the B2CC may be explained partly by the large south eddy that also collected fish. At first, the south eddy appeared to compete with the B2CC entrance for fish, but the eddy collects fish that missed the B2CC entrance on the first pass, and it likely circulates them where they have additional opportunities to discover the entrance before they are entrained in turbines.

The zone with >90% entrance probability undoubtedly extended much further to the north along the powerhouse face than the 47 ft estimated by the Markov chain analysis or even the 62 ft that we could sample with the DIDSON. Fish within 30 ft of the powerhouse face, even as far north as one half of the powerhouse length, were swept south toward the B2CC entrance by strong lateral flows along the powerhouse. The probability that these fish would enter the zone of estimation by the Markov chain probably exceeds 90% also. A capture velocity extending well upstream of the sill was instrumental in taking smolts that otherwise might have initially avoided that outlet.

At B1 Sluice 3C during 2005, over 46 million DIDSON fish images were autotracked during data processing for the FEZ objective. These individual frames made up 2,078,132 tracks (i.e., swim paths) of fish. The majority of these—2,009,022—were tracked during spring data collection; 1,033,566 during the day and 975,456 at night. During summer, 69,110 fish were tracked; 43,964 during the day and 25,146 at night.

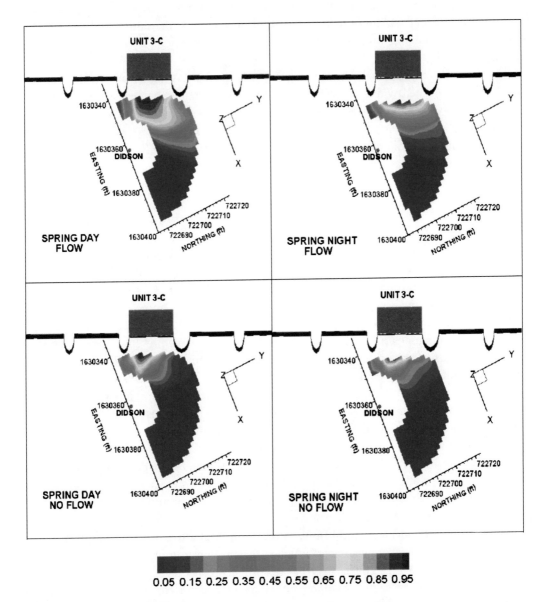

Figure 23. Contours of Fish Passage Probabilities at the B1 Sluice 3C in Spring (top row) and Summer (bottom row) 2005. Probabilities above are shown for the day (left panels) and night (right panels), and powerhouse flow >212 m^3/s (upper panels) and <212 m^3/s (lower panels). Sluiceway passage fates x- and y-scales are in Oregon State Plane feet.

Differences in entrance efficiency are shown in contour plots (Figures 23). The distance associated with 90% entrance efficiency for B1 Sluice 3C was about three times higher in the summer than it was in the spring, and reduced turbine flow in summer may be mostly responsible for observed differences. The average distance associated with the 90% entrance probability for four conditions was about 2.1 m in spring and about 6.6 m in summer. The frequency of sampling hours with <566 m^3/s of powerhouse flow or sluiceway-only flow was much higher in summer than it was in spring, and this would have greatly simplified the flow environment for approaching smolts in summer. There were a few days in spring when thousands of yearling fish were observed holding upstream of the outlet, and this was not observed during summer sampling when the fish approach rate was more consistent and predictable.

The lack of capture velocity to entrain fish at the B1 sluiceway entrances is a shortcoming associated with the limited capacity of the existing channel. Three chain gates could only be opened down to elevation 21.8 m above msl or the channel would be flooded at above-average pool elevations. If the channel had greater capacity, gates could be opened down to elevation 20.7 m above msl, where the existing sill is located. Hydroacoustic data showed that a noticeable percentage of the fish observed moving in an upstream direction over the chain gates at Entrances 1C and 3C also were moving downward in the water column. The chain gates at the B1 sluiceway entrances form a sharp-crested weir at elevation 21.8 m above msl, and they pool a 1.1-m-deep volume of water between the top of the weir and the concrete sill. Flow passing over the weir creates a weak hydraulic roller that fish can use as a flow refuge. From a fish-capture standpoint, it would be much better to open gates to elevation 20.7 m above msl and eliminate any flow refuge that smolts might use to avoid entrainment. Low approach velocities may be more of a hindrance for passing yearlings in spring than for passing subyearlings in summer. Ideally, SFOs from forebays would have capture velocities that extend at least 2 m upstream of the flow control structure. Providing SFOs with an entrainment zone extending upstream of the structure could reduce entrance rejection, decrease forebay residence time and risk of predation, and increase passage of schools of smolts (Ploskey et al. 2006). Flow into the B1 sluiceway outlets was less than one-tenth of the flow into the B2CC. Flow was higher at Entrance 1C near the downstream end of the channel than it was into Entrance 3C or 6C, and hydroacoustic data indicated that Entrance 1C passed the most fish and the highest densities of fish each season (Ploskey et al. 2006). At B1, fish could move above the controlling gate and most of them could still escape by diving into a flow refuge immediately upstream of the chain gate. Predators also used this flow refuge to stage attacks on approaching smolts. Providing surface passage routes without abrupt transitions to the entrainment zone would decrease forebay holding time, risk of predation, and escape from entraining flows.

Biologists must understand local flow, smolt, and predator dynamics on a diel basis to be able to maximize surface-collector performance for any location in a forebay. The forebay location of surface passage routes may be the primary concern to assure that smolts will encounter the outlet because they routinely travel in that area of a forebay, but it should not be the only concern. Based upon the DIDSON fish approach and entrance study in 2005, we recommended eliminating chain gates and the flow refuge they create. We recommended installing a vortex suppression shelf upstream of the sill at sill elevation (20.7 m above msl) to prevent vortices from capturing smolts to the turbine that otherwise would pass into the outlet. We also recommended on-and-off testing of constant lighting at sluiceway entrances at night because smolt schools held up more at night than they did during the day, and there was increased predator activity and success during the night. Of course, eliminating the low velocity water volume immediately upstream of the chain gates would eliminate the flow refuge that predators occasionally use.

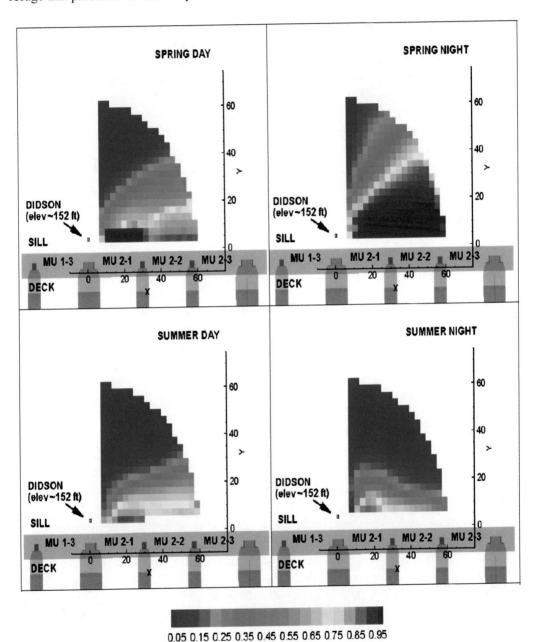

Figure 24. Contours of Fish Passage Probabilities at The Dalles Dam Sluice 2 for Spring and Summer 2005. Sluice fate probabilities are shown for the day (left panels) and night (right panels); x- and y-scales are in feet.

At The Dalles Dam during 2005, a total of 87,762 unique fish were tracked in front of Sluice 2. Of these, 46,902 were present during spring and 40,860 were observed during summer. Figure 24 shows the sums of each spatial cell's fates (states) over the sluiceway absorbing edges after 214 multiplications of the transition matrices. The largest passage probabilities were to the west except for at Sluice 2 during spring night when sluiceway passage was highest. There was little reservoir and east passage. The largest reservoir and east movement occurred during summer night at Sluice 2 (18.6 and 1%). On average, the largest probabilities were to the west (65%) and to the sluiceway (31%). Sluiceway movement varied from 16 to 64% with the lowest values occurring in summer. Sluiceway movement was greatest in spring at Sluice 2. All movement was absorbed at the boundaries and no stagnation occurred.

The FEZ, defined as the point where 90% of fish are entrained, varied from 0 to 5.2 m away from the entrance. The FEZ was spread more to the east than to the west, generally following water flow patterns into the sluiceways (Johnson et al. 2005). The highest sluiceway passage in spring at Sluice 2 corresponded to a FEZ of 2.1 to 5.2 m and is less than the 6-8-m FEZ reported by Johnson et al. (2004) at Sluice 1-1. The FEZ decreased at both sluiceway entrances in the summer study period.

Management Implication

Data on FEZs supported guidelines for entrance conditions at SFOs, including the following: 1) form an extensive SFO flow net (SFO discharge greater than ~7% of total project discharge); 2) create a gradual increase in water velocity approaching the surface flow bypass (ideally, acceleration <1 m/s per meter) so that fish do not avoid an outlet before they are entrained; 3) make water velocities at an entrance high enough (>3 m/s) to capture juvenile fishes in entraining flows near the entrance; and 4) adapt the shape and orientation of the surface entrance(s) to fit site-specific features. These guidelines are being applied during ongoing SFO development at hydroelectric dams on the Columbia and Snake rivers.

FISH AND FLOW RELATIONSHIPS

Fish and flow relationships are statistical associations between fish swimming effort and ambient hydraulic conditions. These relationships provide elemental information on juvenile salmonid responses in flow fields that can be used to support design decisions on SFOs being built to enhance survival rates at mainstem dams on the Snake and Columbia rivers. Sweeney et al. (2007) and Johnson and Dauble (2006) identified concurrent measures of fish movement and hydrodynamics as fundamental research needs for SFO development. The key advance in acoustic imaging for this fisheries research application was merging water velocity and fish velocity data sets. This allowed us to calculate fish swimming effort and use it as the primary dependent variable; fish effort is critical because it is a direct measurement of the direction and speed the fish is swimming.

Background and Objectives

Surface flow outlets are intended to create a flow field in the forebay that juvenile salmon can discover and use to move downstream. Although they generally follow the bulk flow downstream through reservoirs, fish sometimes meander when they encounter slow water in the forebays of dams (Adams et al. 1998). Assuming that smolts discover the SFO flow net, a key point is whether they will react positively or negatively, i.e., will they enter or avoid the entrance? Discovery of a SFO flow net is only part of the issue; another part is for fish to actually follow the flow field and pass into the entrance. Efforts to improve SFO passage led to the spillway weir concepts, but there may be other, less-expensive approaches such as the temporary spillway weirs. To develop these new approaches, basic empirical data on fish response to SFO flow fields is needed to help coalesce engineering design guidelines.

We undertook fish/flow research during 2007 at The Dalles Dam sluiceway (Johnson et al. 2008). The goal of the study was to describe fish behavioral responses to flow fields. We integrated computational fluid dynamic model results with data on smolt movements collected between May 1 and July 12 to address two SFO development objectives: Which hydraulic variables are associated with fish responses? Of these, are there threshold levels that could be used to support SFO design guidelines?

Methods

We collected and merged data from a computational fluid dynamics (CFD) model and a DIDSON. At The Dalles Dam, the DIDSON was deployed upstream off the face of the dam to sample in the near field (<10 m) of Sluices 1-1 and 1-2 (Figure 25). We rotated the apparatus once per day to cover four aiming angles.

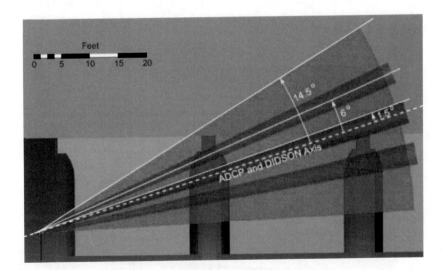

Figure 25. DIDSON Sample Volume – Plan View of DIDSON (purple) Acoustic Beams. The red beams are from an acoustic Doppler current profiler. The background is The Dalles Dam powerhouse. The projection of the sloping piers into the beams is an artifact of the graphic.

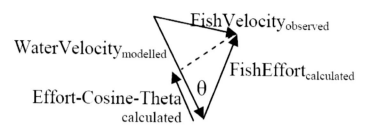

Figure 26. Observed and Calculated Fish Variables.

Hydraulic variables were derived from the CFD simulations of water velocity (30-s running average), including water speed, turbulence index, temporal acceleration index, total acceleration, and total strain index. The strain index is the same as the spatial velocity gradient tensor used to represent total hydraulic strain by Goodwin et al. (2006). Observed fish movement, as measured from the DIDSON images, is the result of the interaction between the flow field, as simulated with CFD, and fish swimming behavior. That is, the observed fish velocity vector is the sum of the water velocity and the fish swimming effort vectors, where theta (θ) is the angle between these two vectors (Figure 26). Thus, fish behavior can be characterized by the following four response variables: 1) fish speed (m/s)—the magnitude of the fish velocity vector; 2) effort speed (m/s)—the magnitude of the fish swimming effort vector; 3) swim angle θ—the angle between the water velocity and fish swimming effort vectors; and 4) effort-cos-theta (m/s)—the magnitude of the effort vector projected on the water vector.

Results

CFD data show the oblique flow into the sluiceway at The Dalles Dam (Figure 27). Flow abruptly accelerates inside the piers and over the sill at the sluiceway entrances.

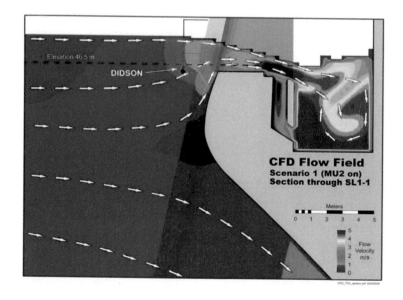

a.

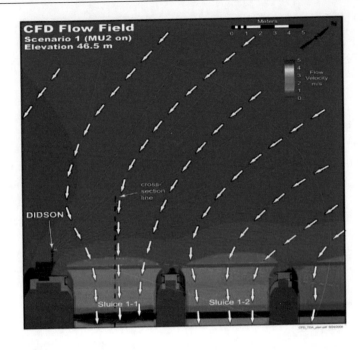

b.

Figure 27. CFD Results in Cross-Sectional (a) and Plan (b) Views Show Abruptly Changing Flow into the Sluiceway at The Dalles Dam, El. 48.3 m above msl. (Total discharge 7,730, spillway 3,115, powerhouse 4,616, Sluice 1-1 and 1-2 77, MU1 280 and MU2 278 m3/s.)

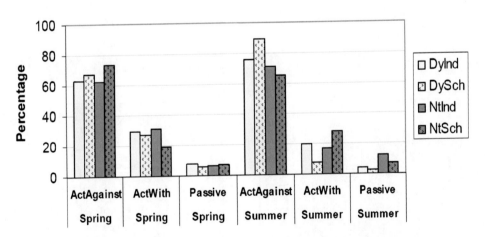

Figure 28. The Most Common Fish Behavior Relative to Flow Was Actively Swimming Against the Flow. Percentages based on effort-cos-theta were calculated seasonally for individual fish and schools during day and night separately; e.g., for spring/day/individuals, the sum of percentages for active against, active with, and passive equals 100.

Fish swimming relative to flow, based on effort-cos-theta to categorize fish behaviors, was 1) passive, 2) active swimming against the flow (positive rheotaxis), and 3) active swimming with the flow (negative rheotaxis). Passive behavior was defined as being within 0.03 m/s of zero; i.e., about one-fifth of a body length per second. The majority behavior was active swimming against the flow (65 to 85%) (Figure 28). Conversely, approximately 10 to

30% of the behavior at The Dalles Dam was active swimming with the flow (negative rheotaxis). A small fraction of swimming behavior was passive (~5%). Swimming against the flow (positive rheotaxis) was more common in summer than spring at The Dalles Dam. Generally, individual fish were less likely to swim against the flow than were schools of fish.

Fish effort superimposed on flow conditions shows relatively high fish-swim-effort values and negative effort-cos-theta just upstream of the sluice entrances (Figure 29). Water velocity increases in this region, as does acceleration and strain.

A correlation analysis shows that effort-cos-theta had higher correlations with hydraulic variables than did fish-swim-effort. The highest correlations (0.46-0.47) were between effort-cos-theta and water velocity magnitude, V (water speed's y-component, perpendicular to the dam), W (water speed's vertical component), total acceleration, and strain. Most spatial derivatives of velocity were not strongly correlated with the fish behavior variables.

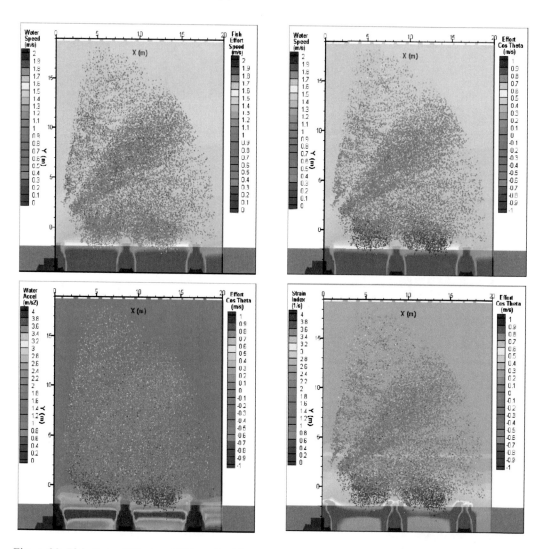

Figure 29. Fish-Swim-Effort and Effort-Cos-Theta Are Associated with Water Velocity Fields (top row), Acceleration Field (bottom left), and Strain Field (bottom right). Hydraulic data are from the CFD simulation. The fish data points are ping-to-ping observations processed from DIDSON output.

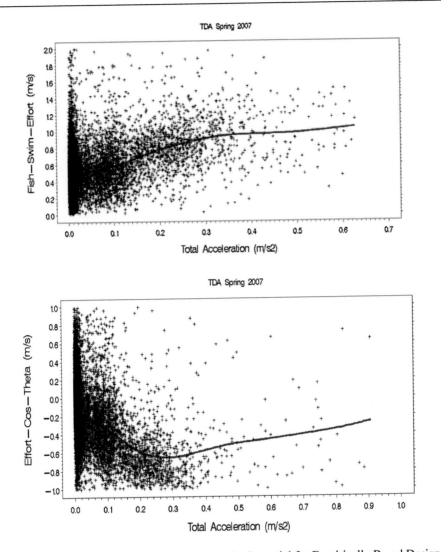

Figure 30. Example Fish/Flow Relationships Indicate the Potential for Empirically Based Design Guidelines. Leveling of the effort variables could indicate a response threshold. Data are for The Dalles Dam, spring 2007, fish-swim-effort (top) and effort-cos-theta (bottom) vs. total acceleration.

A non-linear regression analysis was applied to examine quantitative relationships between the fish behavior variables and hydraulic variables to assess its usefulness to support development of SFO design guidelines. For fish-swim-effort and effort-cos-theta as the dependent variables (Figure 30), the scatter cloud of data points was oriented in upward and downward directions, respectively, as the independent variable increased from its low values during both spring and summer. The corresponding splines reflected this as fish-swim-effort and effort-cos-theta trended upward and downward, respectively, as velocity, acceleration, or strain increased. As an example, during spring, effort-cos-theta peaked at approximate velocity 0.9 m/s, acceleration 0.25 m/s^2, and strain 0.95 s^{-1}. Note that the data were sparse at the high end for all independent variables.

The 2007 results provide the following new information:

1. Schooling behavior was dynamic and prevalent. The implication is that the SFO entrance area must be large enough to accommodate fish schools.
2. Fish behavior was dependent on the distance from the SFO entrance. This supports the notion that SFO flow nets need to be expansive enough spatially to attract smolts despite competing flow fields.
3. Passive fish behavior was observed less than 5% of the time in the SFO flow nets we studied, implying that SFO designs cannot rely only on fish following bulk flow.
4. Active swimming against the flow was the most common behavioral response. Evaluations of SFO performance should include metrics for fish swimming behavior in SFO flow fields.
5. Fish effort variables were correlated with water velocity, acceleration, and strain. The non-linear regressions indicate potential for this approach of merging fish/flow data to lead to SFO design guidelines in the future as the fish/flow data set is further populated.

Management Implication

Statistical associations between juvenile salmonid movements and hydraulic conditions immediately upstream of the sluiceway SFO entrances showed that water speed, strain, and total acceleration were important variables in explaining the variation in fish behavior data. The results indicate the potential for this approach of merging water/fish measurements to lead to SFO design guidelines in the future as the fish/flow data set is populated from a diversity of SFO sites. Analyzing merged fish/flow data from a diversity of sites in multiple years will strengthen the relationships between smolt responses and hydrodynamic conditions such that universal trends may emerge to support bioengineering efforts aimed at protecting juvenile salmonids.

CONCLUSION

This chapter reported acoustic imaging research for four main topics supporting fisheries management decisions to protect juvenile salmonids migrating downstream through hydroelectric facilities on the Columbia River. First, we used the DIDSON to observe *fish behavior* at The Dalles Dam in 2002, and found that fish can hold in front of turbine intake trashracks, predators congregate in low-velocity zones next to pier noses and adjacent to the sills at sluiceway entrances, and predator abundance is positively correlated with smolt abundance (Johnson et al. 2003). Second, at the B2 in 2003 and 2004, we estimated *fish passage rates* of juvenile salmonids into the gap at an intake guidance screen to quantify how many fish passed there and to show that the gap-closure devices designed to prevent that loss were working (Ploskey et al. 2004; Ploskey et al. 2005). Third, analysis of *fish entrainment zones* determined from DIDSON data for sluiceways at B2 during 2004, B1 during 2005, and The Dalles Dam during 2005 showed that the zones were generally larger during night than during day, larger during spring than during summer, and larger with high (144 m^3/s) rather than low (43 m^3/s) inflow (Ploskey et al. 2005; Ploskey et al. 2006; Johnson et al. 2006). The data supported recommendations based on professional judgment for design guidelines for SFO entrance conditions. And, fourth, *fish/flow relationships* used merged DIDSON and CFD data collected at The Dalles Dam during 2007 to calculate the fish swimming effort vector, associate it with hydraulic variables, and identify water-velocity magnitude, strain, and total acceleration as key variables to prioritize in the ongoing effort to establish engineering design guidelines for fish passage structures.

Research on juvenile salmonids using acoustic imaging at Columbia River dams required a series of technological advancements. The first advance was to devise methods to deploy the equipment at dams. We attached the DIDSON to a two-axis rotator that was mounted on a frame and lowered down the face of the dam to an underwater location. The DIDSON/rotator was used in the early fish behavior research and continues to be a useful combination. The next major developments were filtering algorithms and manual tracking software to process the image files. Subsequently, automated tracking software allowed us to increase sample sizes to obtain the number of observations of fish movement necessary to estimate passage rates and entrainment zones. The newest advance reported in this chapter is the collection of acoustic image data on fish movements and CFD simulation data on water velocity to determine fish/flow relationships. Future improvements in acoustic imaging data collection, current measurements, processing, and analysis for fisheries research might include three-dimensional data, faster frame rates (>20 frames/s) at the 1.8-MHz operating frequency, pattern recognition software to identify species, tighter vertical resolution, and a scanning acoustic Doppler velocimeter to sample inhomogeneous flow. Numerous, potential future applications of acoustic imaging exist for fisheries research at hydroelectric facilities. A next-generation DIDSON with increased processing speed deployed inside a turbine intake to sample fish immediately upstream of the stay vane/wicket gate area could determine fish distribution to aid the design of new, fish-friendly turbines.

Additional work on fish behavior near turbine intake trashracks is warranted, perhaps in conjunction with trashracks designed to pass water efficiently for power production while deterring fish passage. For example, at Richard B. Russell Dam on the Savannah River, a bar screen veneer of 0.32-cm-wide wedge wire on 5.08-cm centers (4.8-cm x 15.2-cm openings) was placed directly over the trash racks to divert fishes larger than about 35 cm in length from trashracks (Nestler et al. 1995); this structural modification successfully reduced entrainment of fish >216 mm long into turbines (Ploskey et al. 1995).

The research on fish/flow relationships might be expanded to increase the diversity of dams surveyed to build a more robust data set than is currently available. Acoustic imaging of fish behavior on approach and movement around behavioral guidance structures in forebays (Adams et al. 1998) would be a useful addition to the evaluations. The concept of a portable sluiceway weir at The Dalles Dam could be evaluated using acoustic imaging to compare fish behavior, passage rates, entrainment zones, and fish/flow relationships at unmodified and modified (specially shaped) weirs. New research is needed on predator/prey interactions at dam structures, forebays, and outfalls, as well as on the behavior of adult salmonids at fishway entrances and SFOs. Fisheries and operations managers have been and will continue to use data from acoustic imaging and other monitoring techniques during decision-making to reduce the impacts of hydroelectric dams on fishes.

ACKNOWLEDGEMENTS

U.S. Army Corps of Engineers, Northwestern Division, Portland and Walla Walla Districts directed and funded this research under the Anadromous Fish Evaluation Program using Congressional appropriations for the Columbia River Fish Mitigation Project. The hard work and dedication of the following USACE fisheries biologists, hydraulic engineers, project operations and maintenance personnel made this research possible: Bob Cordie, Dan Feil, Blaine Ebberts, Laurie Ebner, Dick Harrison, Bernard Klatte, Art Kunigel, Mike Langeslay, Rock Peters, Lynn Reese, Dennis Schwartz, Ann Setter, Marvin Shutters, Bob Wertheimer, and Miro Zyndol. Colleagues at the Pacific Northwest National Laboratory helped tremendously in all phases of the work: Al Garcia, Terri Gilbride, James Hughes, Julie Hughes, Kathy Lavender, Russell Moursund, Nate Phillips, Cindy Rakowski, John Serkowski, Jan Slater, Scott Titzler, and Shon Zimmerman. We very much appreciate the mechanical and technical assistance provided by Bob Austin and Mike Honald (Honald Crane Services), Ed Belcher and Bill Hanot (University of Washington, Applied Physics Laboratory),Vinnie Schlosser and staff (Schlosser Machine Shop), John Skalski (University of Washington), John Steinbeck (Tenera Environmental), and Kyle Bouchard, Mike Hanks, Peter Johnson, Jina Kim, Deborah Paterson, and Carl Schilt (BAE). We thank Susan Ennor for editing and Dennis Dauble for reviewing the manuscript.

REFERENCES

Adams, N. S., Evans, S. D., and Rondorf, D. W. 1998. Migrational characteristics of juvenile fall chinook salmon in the forebay of Lower Granite Dam relative to the 1997 surface bypass collection tests. Contract E-86930151 report of the Biological Resources Division, U. S. Geological Survey to the U. S. Army Corps of Engineers, Walla Walla, Washington.

Adams, N., Johnson, G., Rondorf, D., Anglea, S., and Wik, T. 2001. Biological evaluation of the behavioral guidance structure at Lower Granite Dam on the Snake River, Washington in 1998. In: Coutant, C., (Ed.), *Behavioral Technologies for Fish Guidance* (pp. 145-160). Bethesda, Maryland, American Fisheries Society.

Beamesderfer, R. and Nigro, T. 1992. Status and Habitat Requirements of the White Sturgeon Populations in the Columbia River Downstream of McNary Dam. Final Report of Research, Volume I, July 1986 - September 1992. U.S. Department of Energy, Bonneville Power Administration, Division of Fish and Wildlife, P.O. Box 3621, Portland, Oregon.

Belcher, E. O. and Lynn, D. C. 2000. Acoustic, near-video-quality images for work in turbid water. In: Proceedings of Underwater Intervention Conference, January 2000, Houston, Texas.

Belcher, E. O., Dinh, H. Q., Lynn, D. C., and Laughlin, T. J. 1999. Beamforming and imaging with acoustic lenses in small, high-frequency sonars. In: Proceedings of Oceans '99 Conference, September 1999, Seattle, Washington.

Berggren, T. J. and Filardo, M. J. 1993. An analysis of variables influencing the migration of juvenile salmonids in the Columbia River basin. *North American Journal of Fisheries Management 13*:48-63.

Boswell, K. M., Wilson, M. P., and Cowan, J. H. 2008. A semiautomated approach to estimating fish size, abundance, and behavior from dual-frequency identification sonar (DIDSON) data. *North American Journal of Fisheries Management 28*:799–807

Bottom D. L, Simenstad, C. A., Burke, J., Baptista, A. M., Jay, D. A., Jones, K. K., Casillas, E., and Schiewe, M. H. 2005. Salmon at River's End: the Role of the Estuary in the Decline and Recovery of Columbia River Salmon. National Oceanic and Atmospheric Administration (NOAA) Technical Memorandum. NMFS-NWFSC-68, NOAA National Marine Fisheries Service, Seattle, Washington.

Brookner, E. 1998. *Tracking and Kalman Filtering Made Easy*. New York, John Wiley and Sons.

Coutant, C. C. and Whitney, R. R. 2000. Fish behavior in relation to passage through hydropower turbines: a review. *Transactions of the American Fisheries Society 129*:351-380.

Gessel, M. H., Williams, J. G., Brege, D. A., and Krcma, R. F. 1991. Juvenile salmonid guidance at the Bonneville Dam Second Powerhouse, Columbia River, 1983-1989. *North American Journal of Fisheries Management 11*:400-412.

Goodwin, R. A., Nestler, J. M., Anderson, J. J., Weber, L. J., and Loucks, D. P. 2006. Forecasting 3-D fish movement behavior using a Eulerian-Lagrangian agent method (ELAM). *Ecological Modeling 192*:197-223.

Haro, A., Richkus, W., Whalen, K., Hoar, A., Busch, W., Lary, S., Brush, T., and Dixon, D. 2000. Population decline of the American eel: implications for research and management. *Fisheries 25*:7-16.

Hedgepeth, J., Johnson, G., Skalski, J., and Burczynski, J. 2002a. Active fish tracking sonar (AFTS) for assessing fish behavior. *Acta Acustica 88*:739-742.

Hedgepeth, J. B., Johnson, G. E., Giorgi, A. E., and Skalski, J. R.. 2002b. Sonar Tracker Evaluation of Fish Movements Relative to J-Occlusions at The Dalles Dam in 2001. Final report submitted to U.S. Army Corps of Engineers, Portland District by Battelle Pacific Northwest Division, Richland, Washington.

Holmes, J. A., Cronkite, G. M. W., Enzenhofer,H. J., and Mulligan, T. J. 2006. Accuracy and precision of fish count data from a 'dual-frequency identification sonar' (DIDSON) imaging system. *ICES Journal of marine Science 63*:543-555.

Johnson, G. E. and Dauble, D. D. 2006. Surface flow outlets to protect juvenile salmonids passing through hydropower dams. *Reviews in Fisheries Science 14*:213-244.

Johnson, G. E., Hanks, M. E., Hedgepeth, J. B., McFadden, B. D., Moursund, R. A., Mueller, R. P., and Skalski, J. R. 2003. Hydroacoustic Evaluation of Turbine Intake J-Occlusions at The Dalles Dam in 2002. PNWD-3226, Battelle Pacific Northwest Division, Richland, Washington.

Johnson, G. E., Hedgepeth, J. B., Skalski, J. R., and Giorgi, A. E. 2004. A Markov chain analysis of fish movement to determine entrainment zones. *Fisheries Research 69*:349-358.

Johnson, G. E, Hanks, M. E., Khan, F., Cook, C. B ., Hedgepeth, J. B., Mueller, R. P., Rakowski, C. L., Richmond, M. C., Sargeant, S. L., Serkowski, J. A., and Skalski, J. R.

2005. Hydroacoustic Evaluation of Juvenile Salmonid Passage at The Dalles Dam in 2004. PNNL-15180, Pacific Northwest National Laboratory, Richland, Washington.

Johnson, G., Beeman, J., Duran, I., and Puls, A. 2007. Synthesis of Juvenile Salmonid Passage Studies at The Dalles Dam, Volume II: 2001-2005. PNNL-16443, Pacific Northwest National Laboratory, Richland, Washington.

Johnson, G. E, Richmond, M. C., Hedgepeth, J. B., Deng, Z., Khan, F., Mueller, R. P., Steinbeck, J. R., Sather, N. K., Anderson, M. G., and Ploskey, G. R.. 2008. Smolt Responses to Hydrodynamic Conditions in Forebay Flow Nets of Surface Flow Outlets, 2007. PNNL-17387, Pacific Northwest National Laboratory, Richland, Washington.

Keefer, M. L., Peery, C. A., Bjornn, T. C., Jepson, M. A., and Stuehrenberg, L. C. 2004. Hydrosystem, dam, and reservoir passage rates of adult Chinook salmon and steelhead in the Columbia and Snake rivers. *Transactions of the American Fisheries Society 133*:1413–1439.

Kemeny, J. G. and Snell, J. L. 1960. *Finite Markov Chains*. Princeton, New Jersey. D. Van Nostrand Company, Inc.

Mathur, D., Heisey, P. G., Euston, E. T., Skalski, J. R., and Hays, S.. 1996. Turbine passage survival estimation for chinook salmon smolts (*Oncorhynchus tshawytscha*) at a large dam on the Columbia River. *Canadian Journal of Fisheries and Aquatic Science 53*:542-549.

Maxwell, S. L. and Smith, A. V. 2007. Generating river bottom profiles with a dual-frequency identification sonar (DIDSON). *North American Journal of Fisheries Management 27*:1294–1309.

Maxwell, S. L. and Gove, N. E. 2004. The Feasibility of Estimating Migrating Salmon Passage Rates in Turbid Rivers Using a Dual Frequency Identification Sonar (DIDSON) 2002. Regional Information Report1 No. 2A04-05. Alaska Department of Fish and Game, Division of Commercial Fisheries, Central Region, Anchorage, Alaska.

Maxwell, S. L. and Gove, N. E. 2007. Assessing a dual-frequency identification sonars' fish-counting accuracy, precision, and turbid river range capability. *Journal of the Acoustical Society of America 122*:3364-3377.

Michimoto, R. 1971. Bonneville and The Dalles Dams Ice-Trash Sluiceway Studies, 1971. Fisheries Engineering Research Program, Report #20, submitted to the U.S. Army Corps of Engineers – Portland District, by the Oregon Department of Fish and Wildlife, Portland, Oregon.

Moursund, R. A., Dauble, D. D., Langeslay, M. J. 2001. Turbine intake diversion screens: investigating effects on Pacific lamprey. *Hydro Review 22*:40-46.

Moursund, R. A., Carlson, T. J., and Peters, R. D. 2003. A fisheries application of the dual-frequency identification sonar acoustic camera. *ICES Journal of Marine Science 60*:678-683.

Mueller, R. P., Brown, R. S., Hop. H., and Moulton, L. 2006. Video and acoustic camera techniques for studying fish under ice: a review and comparison. *Rev. Fish. Biol. Fisheries 16*:213–226.

National Research Council. 1996. *Upstream: Salmon and Society in the Pacific Northwest*. Washington, D.C., National Academy Press.

National Oceanic and Atmospheric Administration. 2008. Consultation on remand for operation of the Federal Columbia River Power System, 11 Bureau of Reclamation projects in the Columbia basin and ESA Section 10(a)(1)(A) permit for juvenile fish transportation program (Revised and reissued pursuant to court order, NWF v. NMFS, Civ. No. CV 01-640-RE [D. Oregon]). NOAA's National Marine Fisheries Service, Northwest Region, Seattle, Washington.

Neraas, L. P. and Spruell, P. 2001. Fragmentation of riverine systems: the genetic effects of dams on bull trout (*Salvelinus confluentus*) in the Clark Fork River system. *Molecular Ecology 10*:1153-1164.

Nestler, J. M., Ploskey, G. R., Schneider, L. T., and Weeks, G. 1995. Development of an operational, full-scale fish protection system at a major pumped-storage hydropower dam. In: Cassidy. J. J. (Ed.), *Waterpower '95* (pp. 152-161), New York, American Society of Civil Engineers.

Northwest Power and Conservation Council. 2000. Columbia River Basin Fish and Wildlife Program: a Multi-Species Approach for Decision Making. Council document 2000-19, Portland, Oregon.

Pease, B. and Green, B. 2007. Observing fish behaviour at tidal floodgates. *Society of Wetland Scientists Bulletin 24*:29–30

Petersen, J. H. 2001. Density, aggregation, and body size of northern pikeminnow preying on juvenile salmonids in a large river. *Journal of Fish Biology 58*:1137–1148.

Petersen, J. H. and Kitchell, J. F. 2001. Climate regimes and water temperature changes in the Columbia River: bioenergetic implications for predators of juvenile salmon. *Canadian Journal of Fisheries and Aquatic Science 58*:1831–1841.

Ploskey, G. R., Nestler, J. M., Weeks, G., and Schilt, C. 1995. Evaluation of an integrated fish-protection system. In: Cassidy. J. J. (Ed.), *Waterpower '95* (pp. 162-171), New York, American Society of Civil Engineers.

Ploskey G., Poe, T., Giorgi, A., and Johnson, G. 2001. Synthesis of Radio Telemetry, Hydroacoustic, and Survival Studies of Juvenile Salmon at The Dalles Dam (1982-2000). PNWD-3131, Final report submitted to the U.S. Army Corps of Engineers, Portland District, by Battelle Pacific Northwest Division, Richland, Washington.

Ploskey, G. R., Weiland, M. A., and Schilt, C. R. 2004. Evaluation of Fish Losses Through Screen Gaps at Modified and Unmodified Intakes of Bonneville Dam Second Powerhouse in 2003. PNNL-14539, Pacific Northwest National Laboratory, Richland, Washington.

Ploskey, G. R., Weiland, M. A., Schilt, C. R., Kim, J., Johnson, P. N., Hanks, M. E., Patterson, D. S., Skalski, J. R., and Hedgepeth, J. B. 2005. Hydroacoustic Evaluation of Fish Passage Through Bonneville Dam in 2004. PNNL-15249, Pacific Northwest National Laboratory, Richland, Washington.

Ploskey, G. R., Weiland, M. A., Zimmerman, S. A., Hughes, J. S., Bouchard, K., Fischer, E. S., Schilt, C. R., Hanks, M. E., Kim, J., Skalski, J. R., Hedgepeth J. B., and Nagy, W. T. 2006. Hydroacoustic Evaluation of Fish Passage Through Bonneville Dam in 2005. PNNL-15944, Pacific Northwest National Laboratory, Richland, Washington.

Raymond, H. L. 1968. Migration rates of yearling chinook salmon in relation to flows and impoundments in the Columbia and Snake rivers. *Transactions of the American Fisheries Society 97*:356-359.

Raymond, H. L. 1979. Effects of dams and impoundments on migrations of juvenile chinook salmon and steelhead from the Snake River, 1966 to 1975. *Transactions of the American Fisheries Society 108*:505-529.

Rose, C. S., Stoner, A. W., and Matteson, K.. 2005. Use of high-frequency imaging sonar to observe fish behaviour near baited fishing gears. *Fisheries Research 76*: 291-304.

Sherwood, C. R., Jay, D. A., Harvey, R. B., Hamilton, P. and Simenstad, C. A. 1990. Historical changes in the Columbia River estuary. *Progress in Oceanography 25*:299-352.

Simenstad, C. A., Small, L. F., McIntire, C. D., Jay, D. A., and Sherwood, C. R. 1990. Columbia River estuary studies: An introduction to the estuary, a brief history, and prior studies. *Progress in Oceanography 25*:1-13.

Sweeney, C., Hall, R., Giorgi, A., Miller, M., and Johnson, G. 2007. Surface Bypass Program Comprehensive Review Report. Final report submitted to the U.S. Army Corps of Engineers. ENSR Document No. 09000-399-0409, Environmental Consultants and Engineer, Redmond, Washington.

Taylor, H. and Karlin, S. 1998. *An Introduction to Stochastic Modeling*. Academic Press, San Diego, California.

Tiffan, K. F., Rondorf, D. W., and Skalicky, J. J. 2004. Imaging fall Chinook salmon redds in the Columbia River with a dual-frequency identification sonar. *North American Journal of Fisheries Management 24*:1421–1426.

USGS (United States Geological Survey). 1990. Largest Rivers in the United States, in Discharge, Drainage Area, or Length. Open-File Report 87-242, Reston, Virginia.

Wetzel, R. G. 2001. *Limnology: Lake and River Ecosystems*. New York, Academic Press.

Reviewed by Dr. Dennis D. Dauble, Pacific Northwest National Laboratory, P.O. Box 999, Richland, Washington (509 371 7151)

In: Dams: Impacts, Stability and Design
Editors: Walter P. Hayes and Michael C. Barnes

ISBN 978-1-60692-618-5
© 2009 Nova Science Publishers, Inc.

Chapter 2

EFFECTS OF JET GROUTING ON WETLAND INVERTEBRATES AT MORMON ISLAND AUXILIARY DAM, FOLSOM, CALIFORNIA

S. Mark Nelson and Gregory Reed

Technical Service Center, Bureau of Reclamation, Denver, CO 80225 USA

ABSTRACT

Strengthening of dam foundations is sometimes necessary because of potential seismic activity near the dam. A jet grouting test section was undertaken at the Mormon Island Auxiliary Dam (MIAD) near Sacramento, California during the spring and summer of 2007. Jet grouting is the high pressure injection of cement grout slurry deep into soil layers to mix material with grout in order to stabilize the soil structure. The purpose of the test section was to evaluate the feasibility of jet grouting for the foundation soils and to evaluate construction issues such as waste handling, groundwater monitoring, strength/quality of installed concrete, and biological monitoring.

Monitoring of nearby wetlands occurred because some grout materials affect water chemistry and may cause increased pH of groundwater near grout columns. Changes in local geochemistry, such as pH, may have negative effects on wetland environments and existing biota.

Monitoring was undertaken at a wetland complex near MIAD during the period of jet grouting. Results from pH measurements suggested there were no increases in pH coincident with jet grouting activities, and there were no detected impacts to aquatic invertebrates in wetlands adjacent to MIAD. Overall results suggest that jet grouting was a benign activity when water quality and nearby aquatic invertebrate assemblages were considered.

INTRODUCTION

Mormon Island Auxiliary Dam (MIAD) is a 33-m high earthfill dam that helps to impound the American River and form Folsom Lake near Sacramento, California. Because of the potential for seismic activity in the area, the dam foundation was modified in the 1990s to limit seismic deformations. Additional strengthening of the dam foundation is desired, and it has been suggested that jet grouting might achieve this purpose. This technique involves injecting cement grout into the soil (maximum depth to 21-m) to form a series of grout columns which modify the physical properties of the existing soils. Injection of the material is under high pressures and uses large volumes of cement grout. While most of the excess material flows up to the surface where it is contained, it is possible that some of this alkaline cement compound might impact local geochemistry through the release of calcium hydroxide and cause increases in groundwater/surface water pH and allow for increased alkalinity. Alkalinity is a measure of the ability of water to buffer or resist acidity, while pH is a measurement of whether water is acidic or basic. Often these two measures are closely related. There is some evidence in the literature that placement of alkali materials in soils can alter water quality. Murarka et al. (2002) observed increases in surface water alkalinity in a case where coal ash was used to fill a mine pit in Indiana, and similar processes could occur with jet grouting. The purpose of the jet grouting test section was to investigate the viability of the procedure for producing adequate foundation improvement and to test for potential environmental impacts. Despite the relatively common use of jet grouting we found no examples in the literature of studying impacts to aquatic invertebrate assemblages.

Extremely alkaline groundwater is rare in nature and is most often associated with human activities (Roadcap et al. 2005). Water altered in this manner may have profound effects on environments as observed in wetland complexes along Lake Michigan affected by high pH (Roadcap et al. 2005). Analyses of wetlands impacted by high pH limestone quarry water showed large differences in wetland plants between impacted and reference wetlands (Mayes et al. 2005). Laboratory studies of high pH water have demonstrated varying responses from macroinvertebrates; pH increased to $\geq$ 10 had no discernible impact to the midge, *Chironomus*, while amphipod (*Hyalella azteca*) survival decreased by ca. 50 percent after 4 days of exposure (Yee et al. 2000).

Extensive wetland areas exist immediately adjacent and downslope to the jet grouting operations at MIAD (ca. within 345-m). Local resource agencies as well as Bureau of Reclamation (Reclamation) personnel have expressed concern that some harm to aquatic macroinvertebrate assemblages may occur from high pH water entering the wetlands during jet grouting. This paper reports on surface and groundwater monitoring for pH at sites associated with the wetland area and on aquatic invertebrate monitoring. Water chemistry monitoring was designed to determine if changes in pH at wetland sites occurred during jet grouting. Aquatic macroinvertebrates were sampled both before and during jet grouting to determine whether assemblages were altered in association with grouting activities. To avoid confounding jet grouting impacts with natural changes in macroinvertebrate assemblages over time, reference wetlands were also sampled to control for temporal variation. We assumed that impacts would be acute and observable in a short period of time.

METHODS

The basic study design goal was to estimate changes at the MIAD wetland site resulting from jet grouting operations. It should be noted that this grouting operation is not replicated and therefore all data for this study of jet grouting impacts are psuedo-replicated.

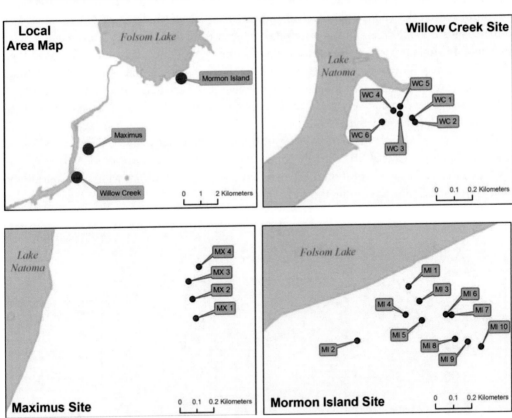

Figure 1. Relative positions of the three groups of sites. MI sites were potentially exposed to grouting impacts, while other sites (MX and WC) were considered to be reference wetlands.

The ability to detect differences caused by an impact at a site may be affected by temporal variation in communities and therefore reference sites were utilized to allow for some confidence in results. It was assumed that temporal variability would be similar between MIAD and reference sites, thus allowing for detection of changes at MIAD from jet grouting impacts. Long-term and frequent water monitoring for pH occurred only at the MIAD site and was used to regulate the grouting operation. The U.S. Environmental Protection Agency's current pH criterion for the protection of freshwater aquatic life (USEPA 1986, 2006) defines an acceptable ambient pH range (i.e., 6.5 to 9.0), and this range was used as a goal during monitoring activities. This criterion does not limit the magnitude of rapid change that organisms can be exposed to within this range, suggesting that the effects of rapid pH changes are insignificant when pH is maintained within the acceptable ambient range. These pH data are provided here as support for conclusions, but cannot be compared to reference sites for discrimination of temporal effects.

Site Locations—The spatial relationship between the three groups of sites [Mormon Island (MI), Maximus (MX), and Willow Creek (WC)] is presented in Figure 1. Sites MI-1 and MI-2, at the MI complex (Figure 1), were closest, and MI-8, MI-9, and MI-10 furthest from the dam. MI-2 and MI-6 were in drainage ditch environments while MI-8, MI-9, and MI-10 appeared to be artificial pools created from historic dam building borrow pits (Sutter and Francisco 1998). Other MI sites were associated with swale areas (MI-1 and MI-7) or vernal pools (MI-3, MI-4, and MI-5). Sites that were given the MX code were found along Willow Creek East of Lake Natoma (Figure 1) and were largely located in the flood plain portion of the creek. There was direct interaction between the lotic creek environment and wetlands lateral to the creek. MX-4 differed somewhat from the other MX sites in that a drain from an industrial property entered the creek at that point, creating a slow moving slough type of environment. Willow Creek sites (Figure 1) were found at the Willow Creek State Recreational Area (Willow Creek SRA), and although they varied in surface area and depth, it appeared that all were relatively small artificial wetlands that may have been associated with historic mining/dredging activity.

Hydrology of the wetland areas likely differed, with flood plain wetlands at MX and swale type environments at MI probably containing water for a large part of the year.

Many of the wetlands at MI and WC may have been vernal, filling with water during the winter rains and drying out during the early summer.

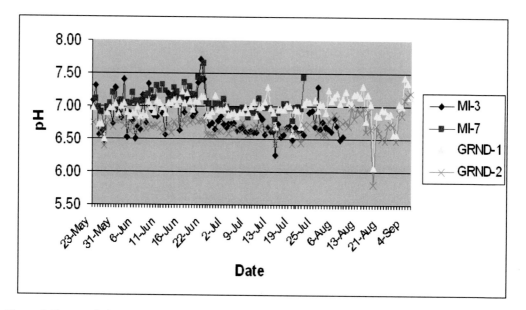

Figure 2. Patterns in long-term pH at surface water (MI-3 and MI-7)) and groundwater (GRND-1 and GRDN-2) MIAD sites.

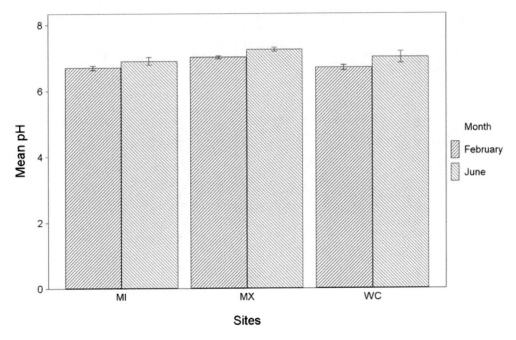

Figure 3. Mean pH values from wetland sampling in February and March. Variance is represented as standard error.

Long-term Monitoring for pH--Monitoring of pH took place at MI-3 and MI-7 surface water locations and at two groundwater wells that were upslope of the wetlands just below MIAD. Sampling took place on a near daily basis starting on May 23, 2007, during jet grouting operations and ended on August 8 for surface water samples (sites were dry at this time) and September 7 for groundwater wells. The pH in groundwater wells was measured at 9 m depth. Meters were calibrated on a weekly basis using the appropriate standards.

Macroinvertebrate Monitoring--Samples were collected before (February 2007) and about 1 month after initiation of jet grouting (June 2007). Samples were collected at 10 sites associated with MIAD that had the potential to be affected by jet grouting and 10 reference sites. Reference sites were palustrine emergent wetlands selected because of gross similarities in vegetation, structure, and flow to MIAD sites, however none of the sites would be considered pristine. Sampling took place at the same 20 sites on each occasion (drought conditions impacted June sampling and resulted in a decreased number of sites) so that they were linked through time. Locations are presented in Figure 1 with the codes MI referring to MIAD associated wetlands and MX and WC referring to reference wetlands near the Maximus building and the Willow Creek SRA.

Aquatic invertebrates were collected with a D-frame net over a 1-minute period. Biotic samples were preserved in alcohol and shipped back to the Technical Service Center laboratory for processing. Other collected variables included dissolved oxygen, conductivity, pH, and temperature as measured with a portable meter. Water samples for alkalinity and hardness were analyzed using titration methods. Estimates of detritus, percent plant cover, sampling depth, vegetation height, and vegetation type were also obtained at each site.

Table 1. Environmental variables measured at wetland sites near Folsom, California

Site	DO (mg/L)-surface	DO (mg/L)-bottom	Cond. (uS/cm)	Temp (°C)	pH	Veg height (m)	Cover (%)	Detritus volume (ml)	Depth (m)	Alkalinity (mg/L)	Hardness (mg/L)
MI-1	3.54-4.99	3.54-4.99	172-175	14.24-16.7	7.02-7.4	1.3-2.0	90	50-400	0.1	62-67	23-24
MI-2	3.85-6.61	3.85-6.61	150-186	8.54-21.01	6.57-7.05	1.5-1.6	100	20-500	0.05	56-70	49-62
MI-3	2.14-5.9	1.22-1.76	169-173	8.9-19.45	6.51-6.74	0.2-0.9	70-100	300-1200	0.5-0.6	63-85	25-28
MI-4[a]	9.4	7.47	216	11	6.9	0.2	50	800	0.2	53	50
MI-5[a]	7.7	7.7	216	12.4	6.92	0.7	60	250	0.1	49	50
MI-6	2.91-5.7	2.91-5.6	174-183	7.6-20.26	6.5-6.98	1.5-2.0	40-75	60-900	0.2-0.3	68-74	28-31
MI-7	1.53-9.14	1.53-9.14	318-360	9.91-24.5	6.6-6.81	3-4	50	125-500	0.05-0.1	89-95	30-52
MI-8	9.47-14.4	7.5-10.2	464-564	9.4-27.7	6.8-7.3	0	0-100	0-900	0.30-0.35	123-142	138-148
MI-9	5.3-12.4	2.36-8.5	389-447	9.8-22.3	6.45-6.6	0	0-50	200-1000	0.35-0.45	126-137	105-137
MI-10	3.4-4.05	3.81-4.15	483-505	9-20.07	6.5-6.6	0-0.1	0-50	500-700	0.3-0.6	147-162	156-183
MX-1	2.3-6.1	2.3-6.1	239-269	10.9-19.6	6.95-7.12	0.8-1.0	80-100	20-125	0.05	87-89	97-143
MX-2	4.41-6.3	4.40-6.3	235-263	10.22-19.88	7.07-7.34	1.5-1.9	85-100	50-250	0.05	85-90	94-101
MX-3	5.15-8.13	5.15-8.13	235-262	11.51-19.75	7.12-7.38	1.7-2.0	95-100	125-500	0.05-0.4	87-93	98-99
MX-4	4.06-7.83	5.06-6.23	217-236	12.14-19.32	6.94-7.17	1.4-2.2	90-100	375-750	0.3	72-77	79-81
WC-1	1.64-3.8	1.64-5.3	286-377	9.1-20.04	6.52-6.81	0.0-0.3	30-40	1200-2000	0.2-0.3	107-117	115-117
WC-2	0.28-5.03	0.28-5.03	274-331	12.86-21.47	6.60-6.63	0.0-0.5	70-100	125-1500	0.05-0.2	116-131	119-133
WC-3	0.54-3.51	0.54-2.01	433-531	10.57-20.18	6.61-6.97	0.0-0.6	20-50	200-1000	0.20-0.25	177-206	164-200
WC-4	1.1-1.97	1.1-2.14	436-487	7.62-18.54	6.7-7.06	0.7-2.0	50-75	250-750	0.2-0.4	188-205	175-210
WC-5	2.8-7.3	2.16-7.3	359-405	11.37-21.3	6.67-7.65	0.2-0.6	30-35	10-1000	0.02-0.25	148-178	146-178
WC-6[a]	3.72	3.46	423	9.23	7.09	0	10	1000	0.3	175	177

[a]These sites were only sampled in February because of drying that occurred in June.

Data Analysis for pH Monitoring--Data were presented as means and ranges. There was no long-term pH data collected before the jet grouting activity was initiated. The main goal from these data was to ensure that pH values did not increase to the point where wetland biota were impacted (pH > 9.0).

Data Analysis for Macroinvertebrate Monitoring--Paired *t*-tests were used to determine whether significant differences in taxa richness occurred from before to during jet grouting activities. Sites at Mormon Island (MI-1 to MI-10) were randomly paired with other reference sites in the Folsom area (MX and WC sites) and number of taxa subtracted from the number of taxa at the reference sites. It was assumed that impacted MI sites would have reduced taxa richness after initiation of jet grout treatment and that this treatment would not impact reference sites that were spatially unassociated with jet grouting. A paired *t*-test was then used to compare data before and during grouting. The alternative hypothesis was that differences in site pairs would be higher in February than in June, therefore leading to a value different than zero and suggesting occurrence of a negative impact between the two periods. In some cases richness measures may be ineffective in detecting impacts because of replacement of sensitive species by tolerant species. We therefore also used ordination to determine whether there were changes in macroinvertebrate assemblages that might potentially be associated with grouting impacts. Paired *t*-tests were also used to compare differences in water quality data between months. In these cases individual sites from the two different months were paired.

Ordination techniques were used to examine patterns in the macroinvertebrate data, and to identify physical and chemical variables that were most closely associated with invertebrate distributions. Initial analysis of the macroinvertebrate data set used detrended correspondence analysis, and revealed that the data set had a gradient length > 3, suggesting that unimodal models were appropriate for analysis. Therefore, canonical correspondence analysis (CCA) was used for direct gradient analysis. Faunal data were transformed (square root transformation) before analysis. Forward selection of environmental variables and Monte Carlo permutations were used to determine whether variables exerted a significant effect on invertebrate distributions. If environmental variables were strongly correlated (Pearson correlation, $r \geq 0.6ll$), only a single variable was selected for use in CCA to avoid problems with multicollinearity. Environmental variables were normalized $[(\ln(X+1))$ or arcsin squareroot transformation for percentage data] if the Shapiro-Wilk Test indicated the necessity for this transformation. In the ordination diagram, taxa and sites are represented by geometric points and the environmental variables by arrows. The arrows roughly orient in the direction of maximum variation of the given variable, and the length of the arrow indicates how much influence a given environmental variable has on macroinvertebrate data. Perpendicular lines drawn from an arrow to macroinvertebrate taxon points determine the relative position of that taxon along the environmental axis.

RESULTS

Jet grouting along the downstream toe of MIAD started on May 17, 2007. This initial activity ended on May 28 but was resumed from June 11 to 15 and June 18 to 20. Aquatic macroinvertebrates were collected on February 6 and 7 and then again on June 12 and 13. Sampling of macroinvertebrates in June was a compromise between allowing enough time for potentially impacted water to reach the wetlands and sampling before rapidly drying pools were lost for the season. Final sampling occurred several weeks after jet grouting initiation and during a period while it was still ongoing.

Long-term monitoring of pH suggested that there were no large deleterious increases in pH that occurred while jet grouting. Values for pH at MI-3 averaged 6.86 and ranged from 6.26 to 7.71 (n=90) while those at MI-7 averaged 7.06 and ranged from 6.63 to 7.65 (n=77). Overall, pH in groundwater wells ranged from 5.80 to 7.43. Averages for pH at the two wells were 6.98 (n=113) and 6.69 (n=114), respectively. Figure 2 shows pH data for the sampling period and suggests that values did not even reach a pH of 8 in either surface water (MI-3 and MI-7) or groundwater (GRND-1 and GRND-2) at any time during or following the jet grouting process.

Environmental variables from the various wetland sites are presented in Table 1. There were differences in hardness between groups of sites, with alkalinity and hardness generally lower at MX compared to WC sites (Table 1). Water quality values ranged widely at the MI complex (Table 1). In general, conditions at MX appeared to be more stable than at other sites with, for example, maximum ranges in conductivity at given MI and WC sites of around 100 µS/cm, while the maximum range at MX sites was 30 µS/cm. This also, by and large, appeared to be the case for alkalinity and hardness measurements. Others have noted that concentrations of dissolved substances vary more widely in temporary waters than in most permanent waters (Williams 2005) such as the perennial flood plain sites at MX. With the exception of MX sites, all sites decreased in wetted area (pers. obs.) between February and June. Three sites were totally without water, MI-4, MI-5, and WC-6. Paired t-tests with pH data from February and June at MI sites indicated that there were significant differences in pH between the two dates (T=-2.89, p=0.0233). However, this was also the case with reference sites (T=-3.52, p=0.0078). It appeared that pH increased at all sites in June when compared to February (Figure 3).

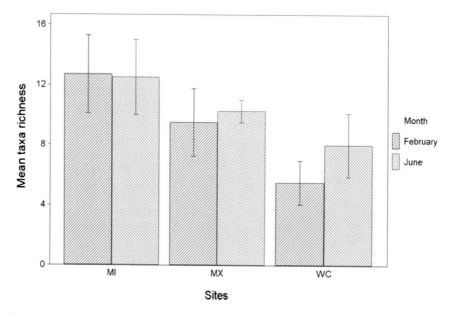

Figure 4. Mean taxa richness at groups of sites in February and June. Variance is represented as standard error.

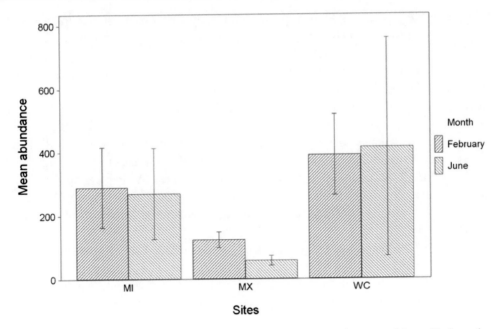

Figure 5. Mean invertebrate abundance at groups of wetland sites in February and June. Variance is represented as standard error.

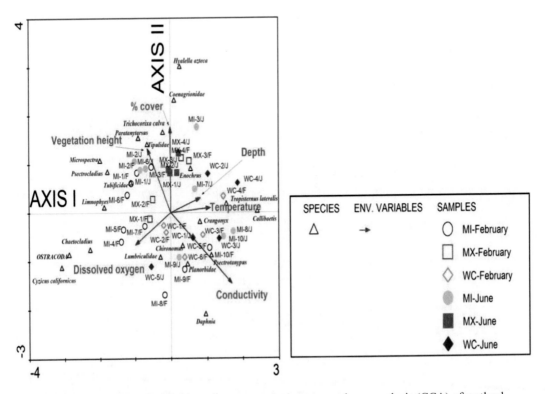

Figure 6. Taxon conditional triplot based on a canonical correspondence analysis (CCA) of wetland macroinvertebrate data with respect to environmental variables. Taxa are represented as small triangles. Only taxa whose fit to the diagram is > 5 percent are shown.

Vegetation such as cattails (*Typha* sp.) were found at a most of the sites. Other emergent plants such as bulrush (*Schoenoplectus* sp.), sedges (*Cyperus* sp.), and spikerush (*Eleocharis* sp.) were more commonly observed at MX and WC locations than they were at MI sites.

Folsom wetland sites contained 105 different macroinvertebrate taxa during the two sampling periods (Appendix A). Only sites that contained water in June were used in the taxa richness analysis, resulting in a decrease to eight pairs of sites used in the analysis. Sites from reference and MI locations were randomly paired together and differences in taxa richness between the pairs calculated for February and June. The 1-tailed test resulted in a P value of 0.5592 (T= -0.15) suggesting that there were no pairwise differences between divergences in taxa richness values before and during grouting at the MI sites. A simple graph comparing mean values at each group of sites supports the paired *t*-test results (Figure 4). We also present a graph of abundance which suggests similar abundance at MI sites between the two periods (Figure 5).

Results of CCA analysis had eigenvalues of 0.436 and 0.367 for the first two axes and explained 13.7 percent of the species data variation and 47.6 percent of the species-environment relation. Initial environmental variables used in the model included percent cover, conductivity, depth, dissolved oxygen at the surface, detritus, pH, temperature, and vegetation height. Other variables were not used because of high ($r>0.6$, $p<0.05$) significant correlations with other model variables. Variables found to be significant in the model were dissolved oxygen, percent cover, conductivity, depth, vegetation height, and temperature. The weighted correlation matrix showing the relationship between species axes and significant environmental variables is presented in Table 2. The water quality indicator of concern in this study, pH, was not a significant variable ($p=0.1738$) in this model. Water temperature and depth were largely associated with the positive portion of the first axis (Figure 6, Table 2) (water temperature was most highly correlated with Axis 3 which is not presented in the figure). Dissolved oxygen was negatively associated with Axis 1 (Figure 6). Conductivity was almost equally weighted along Axis I and Axis II (Table 2).

Paired *t*-tests suggested that water temperature changed with season at sites, and there were significant differences between February and June (T= -13.20, $p<0.0001$) temperatures. Average February temperatures were $10.3 \pm$ SE $0.4\,^{\circ}$C, while in June temperatures averaged $20.7 \pm$ SE $0.6\,^{\circ}$C. Other variables associated with the positive portion of Axis 1 did not appear to differ with sampling occasion, and no significant differences were found using paired *t*-tests (depth, T=2.06, $p=0.0558$ and conductivity, T=0.31, $p=0.7620$). These two variables were likely associated with intrinsic site differences rather than season.

Table 2. Weighted correlation matrix showing relationship between species axes and environmental variables

Variable	Axis			
	1	2	3	4
Dissolved oxygen-surface	**-0.3545**	-0.2663	-0.2791	0.1978
% cover	-0.0040	**0.7088**	0.0152	0.4326
Conductivity	**0.6086**	-0.5880	0.0598	-0.1099
Depth	**0.2931**	0.1214	-0.2674	-0.0572
Vegetation height	-0.2332	**0.5290**	0.0202	-0.4351
Water temperature	0.3929	0.0394	**-0.4809**	0.3002

High correlations associated with a given variable are shown in bold.

Organisms that appeared to be associated with seasonal patterns included *Cyzicus californicus*, an endemic crustacean associated with vernal pools and found in Figure 6 in the negative portion of Axis I. This species and members of the Class Ostracoda were mostly associated with MI-4 and MI-5 locations that were dried by the second sampling occasion in June. *Tropisternus lateralis* was associated with the positive portion of Axis I, and abundance of this single generation beetle has been found to peak in June in other California studies (Zalom et al. 1980). The mayfly, *Callibaetis*, has also been found to increase in abundance in ephemeral ponds over time (Moorhead et al. 1998), and this may partially explain its presence in the extremely positive portion of Axis I. Several midges (e.g., *Chaetocladius*, *Limnophyes*, *Micropsectra*, *Psectrocladius*, and *Psectrotanypus*) appeared to be important in the diagram and may be associated with hydroperiod. In a study of prairie ponds, Driver (1977) found *Pscetrotanypus* to be characteristic of semi-permanent ponds, similar to those in the present study. This genus was most common at MI-7, MI-8, MI-9, WC-1, and WC-5 in June. It appeared that seasonal differences were related both to wetlands having different hydroperiods and the natural history phenology of some invertebrates. The documented (Yee et al. 2000) high-pH sensitive organism, *Hyalella azteca*, was present in MI-3 and MI-6 in February (4 and 13 individuals, respectively) and in MI-3, MI-6, and MI-7 (90, 2, and 2 individuals, respectively) in June, also providing evidence that differences in assemblages between sampling periods were caused by something other than jet grouting impacts.

The second CCA axis appeared to be largely vegetation associated, with percent cover and vegetation height located high in the positive portion of Axis II (Figure 6). *Hyalella azteca* was located in Figure 6 high on Axis II. Edwards and Cowell (1992) found *H. azteca* densities associated with larger emergent macrophyte densities in a Florida lake and suggested that there were greater amphipod food resources and refuge from fish predators at these locations. In the present study it appeared that highest abundance of *H. azteca* was found at sites with highest percent cover which was sometimes associated with emergent vegetation. *Daphnia* were in the highly negative portion of Axis II (Figure 6) and, in some studies; this genus has been found to prefer more pelagic environments and to avoid areas with dense plants (Meerhoff et al. 2006). Burks et al. (2001) also found decreased *Daphnia* densities with increasing macrophyte density, and reported a relationship between the presence of odonates and declines in *Daphnia* density. Odonates in the family Coenagrionidae were found in the CCA diagram high on Axis II and were associated with increased vegetation height and percent cover and were opposite of the sites containing the highest abundance of *Daphnia*. There also appeared to be a difference in wetland types along Axis II with flood plain wetlands at MX sites along with drainage ditch and swale-type wetlands (MI-1, MI-2, and MI-6) found high on Axis II while the more semi-permanent pond sites at MI and WC were found low on Axis II in association with higher conductivities.

Sites from the various locations tended to be mixed together, although there was some tendency for MX sites to cluster together. It did not appear that differences in macroinvertebrate assemblages along either Axis I or Axis II were associated with grouting activities and pH was not a significant variable in the model.

CONCLUSION

Construction activities involving jet grouting did not seem to impact either surface or groundwater pH chemistry. Increased pH values from February to June appeared to be related to temporal patterns that occurred regionally. In a study of the impact of grouting on groundwater in the City of Berlin, Eiswirth et al. (1999) detected no significant changes in groundwater chemistry. This was in a highly porous aquifer, and it was suggested that NaOH, which leached out of the grout, quickly became immobile through buffering reactions with the groundwater. Impacts of jet grouting on water chemistry at MIAD also appear to be limited in extent.

While changes in macroinvertebrate taxa richness were not detected, there were some changes in the makeup of macroinvertebrate assemblages between sampling periods; however, these were likely associated with natural seasonal changes, and pH was not a significant variable in the CCA model. Aside from seasonal changes, macroinvertebrate assemblages, in large part, were associated with wetland characteristics that were independent of jet grouting activities such as dissolved oxygen, vegetation height, and percent cover. Spatial and temporal variability in wetland macroinvertebrate assemblages among the 20 study sites was likely unaffected by jet grouting, but instead merely a response to natural environmental factors along with life history patterns. High alkalinity wetlands in Wisconsin were found to contain characteristic taxa which included isopods, physid snails, and pygmy backswimmers (Pleidae) (Lillie 2003). This sort of assemblage was not found in wetlands at the MIAD and along with the presence of taxa sensitive to high pH suggests a lack of jet grouting impacts.

We believe that the absence of a detectable effect of jet grouting on macroinvertebrates likely reflects the lack of short-term biologically meaningful impacts to this nearby potential environmental stressor. It might be anticipated that some wetland types would be unaffected by altered groundwater chemistry. Vernal wetlands might not be expected to be affected by localized groundwater impacts because of the typically limited exchange with those water sources (e.g., Hanes and Stromberg 1998). Direct precipitation is usually the dominant source of water in this type of wetland (Hanes and Stromberg 1998). However, jet grouting with large amounts of water and pressure could result in the development of a hydraulic gradient that moves toward wetlands that are lower in the landscape, and there are a variety of types of wetlands at MIAD. The assumption in this discussion is that changes in macroinvertebrate assemblages could be observed during this initial impact, if one occurred. Further sampling would be needed if it was expected that impacts would occur upon refilling of wetlands during the winter months. Typically, the types of stressors that exhibit lagged responses are long-term stressors that result in habitat fragmentation and restricted movement between populations of species (e.g., Findlay and Bourdages 2000). These sorts of stressors may not be detectable during short timeframes. It seems unlikely that jet grouting, in the way it was used at MIAD, would result in this type of response. However, monitoring should be extended to ensure this is the case.

APPENDIX A

Macroinvertebrates Collected From Folsom Wetlands in February and June 2007

Folsom CA/February 2007 Site	MI-1	MI-2	MI-3	MI-4	MI-5	MI-6	MI-7	MI-8	MI-9	MI-10	MX-1	MX-2	MX-3	MX-4	WC-1	WC-2	WC-3	WC-4	WC-5	WC-6
Date (2007)	2/6	2/6	2/6	2/6	2/6	2/6	2/6	2/6	2/6	2/6	2/7	2/7	2/7	2/7	2/6	2/6	2/6	2/6	2/6	2/6
COLLEMBOLA				1										1						
EPHEMEROPTERA																				
Baetidae																				
Callibaetis sp.														1						
Centroptilum sp.	10																			
Fallceon quilleri																				
ODONATA																				
Aeshnidae																				
Anax sp.												1		6						
Coenagrionidae			58			11					1	7		10						
Libellulidae																				
Argia sp.	3												1							
Erythemis sp.			1																	
Pachydiplax longipennis																				
Plathemis lydia																				
HETEROPTERA																				
Corixidae																				
Corixidae larvae												11		1						
Hesperocorixa laevigata			1	3	1			3												
Sigara sp.					1															
Trichocorixa calva												2								
Gerridae larvae																				
Macroveliidae																				
Macrovelia hornii																				
Mesoveliidae																				
Mesovelia mulsanti																				
Notonectidae																				
Notonecta kirbyi				3																
Notonecta undulata																				
Saldidae																				

Folsom CA/February 2007

Site	MI-1	MI-2	MI-3	MI-4	MI-5	MI-6	MI-7	MI-8	MI-9	MI-10	MX-1	MX-2	MX-3	MX-4	WC-1	WC-2	WC-3	WC-4	WC-5	WC-6
Date (2007)	2/6	2/6	2/6	2/6	2/6	2/6	2/6	2/6	2/6	2/6	2/7	2/7	2/7	2/7	2/6	2/6	2/6	2/6	2/6	2/6
Saldula sp.																				
TRICHOPTERA																				
Limnephilidae																				
Limnephilus sp.			2				6													
COLEOPTERA																				
Dytiscidae																				
Acilius abbreviatus															1					
Agabus sp.				9	8				2											
Colymbetes sp.					3															
Dytiscus sp.					1												1			
Hydroporinae																				
Hydroporus sp.						1	1	1	1											
Liodessus obscurellus																				
Neoclypeodytes ornatellus							9													
Rhantus sp.																				
Sanfilippodytes sp.				4	4										1					
Haliplidae																				
Peltodytes callosus						1														
Hydraenidae																				
Hydraena sp.					1							1								
Hydrophilidae																				
Anacaena sp.																				
Cymbiodyta sp.													1							
Enochrus californicus				1			2			1										
Enochrus sp.																				
Helochares normatus																				
Paracymus sp.															1					
Tropisternus lateralis										1					2					
Scirtidae																				
Cyphon sp.																				
Staphylinidae																				
DIPTERA																				
Ceratopogonidae																				
Ceratopogoninae						3														
Bezzia/Palpomyia																				
Dasyhelea sp.																				
Chironomidae																				

Folsom CA/February 2007

Site	MI-1	MI-2	MI-3	MI-4	MI-5	MI-6	MI-7	MI-8	MI-9	MI-10	MX-1	MX-2	MX-3	MX-4	WC-1	WC-2	WC-3	WC-4	WC-5	WC-6
Date (2007)	2/6	2/6	2/6	2/6	2/6	2/6	2/6	2/6	2/6	2/6	2/7	2/7	2/7	2/7	2/6	2/6	2/6	2/6	2/6	2/6
Orthocladiinae																				
Chaetocladius sp.			2	53	3	1	129													
Corynoneura sp.				3	2															
Cricotopus sp.	2												4							
Eukiefferiella sp.				1																
Limnophyes sp.	2		1	1	2	3	18			1										
Parametriocnemus sp.	8																			
Paraphaenocladius sp.						2														
Psectrocladius sp.	1		40	7	1															
Rheocricotopus sp.						1														
Smittia sp.							1													
Thienemanniella sp.																				
Chironominae																				
Apedilum sp.	1														2					
Chironomus sp.	1		23	8	3	125				1			1	2						
Dicrotendipes sp.						1														
Endochironomus sp.																				
Glyptotendipes sp.													4							
Paratendipes sp.						2							2							
Phaenopsectra sp.													1							
Polypedilum sp.						11														
Pseudochironomini																				
Pseudochironomus sp.																				
Tanytarsini																				
Micropsectra sp.	10		1	1		15	3								1					
Paratanytarsus sp.			6			14							2							
Rheotanytarsus sp.													1							
Tanytarsus sp.	2			1																
Tanypodinae																				
Ablabesmyia sp.						4														
Alotanypus sp.																				
Larsia sp.	2																			
Procladius sp.						5														
Psectrotanypus sp.													2							
Tanypus sp.																				
Zavrelimyia sp.							1													
Culicidae																				

Folsom CA/February 2007

Site	MI-1	MI-2	MI-3	MI-4	MI-5	MI-6	MI-7	MI-8	MI-9	MI-10	MX-1	MX-2	MX-3	MX-4	WC-1	WC-2	WC-3	WC-4	WC-5	WC-6
Date (2007)	2/6	2/6	2/6	2/6	2/6	2/6	2/6	2/6	2/6	2/6	2/7	2/7	2/7	2/7	2/6	2/6	2/6	2/6	2/6	2/6
Anopheles sp.																				
Culex sp.																				
Dixidae																				
Dixella sp.												1								
Ephydridae																				
Simuliidae																				
Simulium sp.	1												1							
Stratiomyidae																				
Odontomyia sp.																				
Tabanidae				2																
Tipulidae	1			1																
Limnophila sp.						2														
Limonia sp.			2											4						
Tipula sp.																				
TURBELLARIA																				
OLIGOCHAETA							1								1	4				
Enchytraeidae						1														
Lumbricidae											1									
Lumbriculidae	3			4		35	6	9	27	14						7				
Naididae															1	9				
Tubificidae	4			3	1	20			3				8			1				
HIRUDINEA																				
Erpobdellidae																				
Glossiphoniidae													1							
Helobdella stagnalis						2														
CLADOCERA																				
Chydoridae																				
Eurycercus sp.																				
Daphniidae																				
Daphnia sp.										4										
Simocephalus sp.				7	16										5			10	9	104
COPEPODA	16			1																
OSTRACODA				897	776															
SPINICAUDATA																				
Cyzicidae																				
Cyzicus californicus				77	140	1														
AMPHIPODA																				

Folsom CA/February 2007

	MI-1	MI-2	MI-3	MI-4	MI-5	MI-6	MI-7	MI-8	MI-9	MI-10	MX-1	MX-2	MX-3	MX-4	WC-1	WC-2	WC-3	WC-4	WC-5	WC-6
Site	MI-1	MI-2	MI-3	MI-4	MI-5	MI-6	MI-7	MI-8	MI-9	MI-10	MX-1	MX-2	MX-3	MX-4	WC-1	WC-2	WC-3	WC-4	WC-5	WC-6
Date (2007)	2/6	2/6	2/6	2/6	2/6	2/6	2/6	2/6	2/6	2/6	2/7	2/7	2/7	2/7	2/6	2/6	2/6	2/6	2/6	2/6
Crangonyctidae											71	119	58	80	934	93	166	326	297	253
Crangonyx sp.																				
Hyalellidae																				
Hyalella azteca			4			13						10	4	70						
DECAPODA																				
Cambaridae	5		1		2	4	4		77	2	1	1		4			1	1	11	2
ACARI																				
Pionidae																				
Piona sp.										1		2								
GASTROPODA																				
Lymnaeidae															54	34				
Physidae		2	6			8	2	3							2	7				
Planorbidae						1														
BIVALVIA																				
Sphaeriidae																				
Sphaerium sp.																	5			
Total number of taxa	17	1	13	22	18	26	13	4	5	8	4	9	15	10	11	9	4	3	3	3
Total number of organisms	72	2	146	1088	967	287	183	16	110	25	74	153	91	179	1003	159	173	337	317	359

Folsom CA/June 2007

	MI-1	MI-2	MI-3	MI-4	MI-5	MI-6	MI-7	MI-8	MI-9	MI-10	MX-1	MX-2	MX-3	MX-4	WC-1	WC-2	WC-3	WC-4	WC-5	WC-6
Site	MI-1	MI-2	MI-3	MI-4	MI-5	MI-6	MI-7	MI-8	MI-9	MI-10	MX-1	MX-2	MX-3	MX-4	WC-1	WC-2	WC-3	WC-4	WC-5	WC-6
			dry	dry	dry															dry
Date (2007)	6/12	6/12	6/12	6/12	6/12	6/12	6/12	6/12	6/12	6/12	6/13	6/13	6/13	6/13	6/13	6/13	6/13	6/13	6/13	6/13
COLLEMBOLA																		1		
EPHEMEROPTERA																				
Baetidae															12		3			
Callibaetis sp.			17				4	63	5	49	8									
Centroptilum sp.																				
Fallceon quilleri	9																			
ODONATA																				
Aeshnidae														1	1					
Anax sp.			1																	

Folsom CA/June 2007

Site	MI-1	MI-2	MI-3	MI-4 (dry)	MI-5 (dry)	MI-6	MI-7	MI-8	MI-9	MI-10	MX-1	MX-2	MX-3	MX-4	WC-1	WC-2	WC-3	WC-4	WC-5	WC-6 (dry)
Date (2007)	6/12	6/12	6/12	6/12	6/12	6/12	6/12	6/12	6/12	6/12	6/13	6/13	6/13	6/13	6/13	6/13	6/13	6/13	6/13	6/13
Coenagrionidae	4		80			1	23			11		2		3						
Argia sp.																				
Libellulidae																				
Erythemis sp.			2																	
Pachydiplax longipennis			4																	
Plathemis lydia													1							
HETEROPTERA																				
Corixidae																				
Corixidae larvae								6				15	1							
Hesperocorixa laevigata						8	4													
Sigara sp.														1						
Trichocorixa calva						2						25	1	6	1				1	
Gerridae larvae															2		1			
Macroveliidae																				
Macrovelia hornii																		1		
Mesoveliidae																				
Mesovelia mulsanti														2						
Notonectidae																				
Notonecta kirbyi																	1			
Notonecta undulata							1													
Saldidae																				
Saldula sp.			1																	
TRICHOPTERA																				
Limnephilidae																				
Limnephilus sp.																				
COLEOPTERA																				

Folsom CA/June 2007

Site	MI-1	MI-2	MI-3	MI-4 (dry)	MI-5 (dry)	MI-6	MI-7	MI-8	MI-9	MI-10	MX-1	MX-2	MX-3	MX-4	WC-1	WC-2	WC-3	WC-4	WC-5	WC-6 (dry)
Date (2007)	6/12	6/12	6/12	6/12	6/12	6/12	6/12	6/12	6/12	6/12	6/13	6/13	6/13	6/13	6/13	6/13	6/13	6/13	6/13	6/13
Dytiscidae																				
Acilius abbreviatus																				
Agabus sp.																				
Colymbetes sp.																				
Dytiscus sp.																				
Hydroporinae	1		1																	
Hydroporus sp.							15								1					
Liodessus obscurellus																				
Neoclypeodytes ornatellus							2													
Rhantus sp.																			1	
Sanfilippodytes sp.			1				3													
Haliplidae																				
Peltodytes callosus																				
Hydraenidae																				
Hydraena sp.							3													
Hydrophilidae																				
Anacaena sp.							3													
Cymbiodyta sp.							11	1												
Enochrus californicus			2				1													
Enochrus sp.	6		4							7										
Helochares normatus												12								
Paracymus sp.			1												65					
Tropisternus lateralis			8				3													
Scirtidae																				
Cyphon sp.	1		1																	

Folsom CA/June 2007

Site	MI-1	MI-2	MI-3	dry MI-4	dry MI-5	MI-6	MI-7	MI-8	MI-9	MI-10	MX-1	MX-2	MX-3	MX-4	WC-1	WC-2	WC-3	WC-4	WC-5	dry WC-6
Date (2007)	6/12	6/12	6/12	6/12	6/12	6/12	6/12	6/12	6/12	6/12	6/13	6/13	6/13	6/13	6/13	6/13	6/13	6/13	6/13	6/13
Staphylinidae			2				2				2	1	1							
DIPTERA																				
Ceratopogonidae																				
Ceratopogoninae										4										
Bezzia/Palpomyia									3											
Dasyhelea sp.										18										
Chironomidae																				
Orthocladiinae																				
Chaetocladius sp.																				
Corynoneura sp.										1	1									
Cricotopus sp.	1																			
Eukiefferiella sp.																				
Limnophyes sp.		1																		
Parametriocnemus sp.	2																			
Paraphaenocladius sp.																				
Psectrocladius sp.									1											
Rheocricotopus sp.																				
Smittia sp.																				
Thienemanniella sp.													1							
Chironominae																				
Apedilum sp.																				
Chironomus sp.						4	103		301	44					786					
Dicrotendipes sp.													1						1	
Endochironomus sp.									1	6										
Glyptotendipes sp.										1										

Folsom CA/June 2007				dry	dry															dry
Site	MI-1	MI-2	MI-3	MI-4	MI-5	MI-6	MI-7	MI-8	MI-9	MI-10	MX-1	MX-2	MX-3	MX-4	WC-1	WC-2	WC-3	WC-4	WC-5	WC-6
Date (2007)	6/12	6/12	6/12	6/12	6/12	6/12	6/12	6/12	6/12	6/12	6/13	6/13	6/13	6/13	6/13	6/13	6/13	6/13	6/13	6/13
Paratendipes sp.													4							
Phaenopsectra sp.																				
Polypedilum sp.		1								2		2								
Pseudochironomini																				
Pseudochironomus sp.										45										
Tanytarsini																				
Micropsectra sp.	16	4											6						1	
Paratanytarsus sp.							1			2	1	1	2							
Rheotanytarsus sp.																	1			
Tanytarsus sp.										1										
Tanypodinae																				
Ablabesmyia sp.										11					5					
Alotanypus sp.							1			36										
Larsia sp.														1						
Procladius sp.						1	44	6	57	3					420	1	3		13	
Psectrotanypus sp.			2																	
Tanypus sp.						1	2													
Zavrelimyia sp.	1																			
Culicidae																				
Anopheles sp.										1		1					10		8	
Culex sp.																				
Dixidae																				
Dixella sp.	1	2																		
Ephydridae										22										
Simuliidae																				
Simulium sp.		1																		
Stratiomyidae																				

Folsom CA/June 2007

Site	MI-1	MI-2	MI-3	MI-4 (dry)	MI-5 (dry)	MI-6	MI-7	MI-8	MI-9	MI-10	MX-1	MX-2	MX-3	MX-4	WC-1	WC-2	WC-3	WC-4	WC-5 (dry)	WC-6 (dry)
Date (2007)	6/12	6/12	6/12	6/12	6/12	6/12	6/12	6/12	6/12	6/12	6/13	6/13	6/13	6/13	6/13	6/13	6/13	6/13	6/13	6/13
Odontomyia sp.										2										
Tabanidae																				
Tipulidae																				
Limnophila sp.	1																			
Limonia sp.	38									10										
Tipula sp.											1									
TURBELLARIA															2					
OLIGOCHAETA																				
Enchytraeidae																				
Lumbricidae											2									
Lumbriculidae						1									7	11				
Naididae									2				1					1		
Tubificidae							2						2							
HIRUDINEA																				
Erpobdellidae																				
Glossiphonidae																				
Helobdella stagnalis							1								6					
CLADOCERA																				
Chydoridae																				
Eurycercus sp.									3		1									
Daphniidae																				
Daphnia sp.									869	5										
Simocephalus sp.									3											
COPEPODA																				
OSTRACODA												1								
SPINICAUDATA													13							
Cyzicidae																				

Folsom CA/June 2007																			
Site	MI-1	MI-2	MI-3	MI-4 (dry)	MI-5 (dry)	MI-6	MI-7	MI-8	MI-9	MI-10	MX-1	MX-2	MX-3	MX-4	WC-1	WC-2	WC-3	WC-4	WC-5	WC-6 (dry)
Date (2007)	6/12	6/12	6/12	6/12	6/12	6/12	6/12	6/12	6/12	6/12	6/13	6/13	6/13	6/13	6/13	6/13	6/13	6/13	6/13	6/13
Cyzicus californicus																				
AMPHIPODA																				
Crangonyctidae																				
Crangonyx sp.											5	21	58	21	342	35	37	20	113	
Hyalellidae																				
Hyalella azteca			90			2	2							10						
DECAPODA																				
Cambaridae												2								
ACARI																				
Pionidae																				
Piona sp.																				
GASTROPODA																				
Lymnaeidae			2				1					3		1	91	7				
Physidae															49					
Planorbidae																				
BIVALVIA																				
Sphaeriidae																				
Sphaerium sp.																				
Total number of taxa	12	5	18	0	0	8	22	4	10	21	8	11	11	11	16	5	8	4	7	0
Total number of organisms	81	9	220	0	0	20	232	76	1245	281	21	74	90	48	1791	66	57	23	138	0

ACKNOWLEDGMENTS

We thank Deb Martin (Entrix) for accompanying us in the field. Sampling was performed under permit with the California Department of Fish and Game (SC-009310). Reclamation's Dam Safety Program provided funding. We'd also like to thank Shawn Oliver (Central California Area Office - Folsom) for his tireless support of the project.

REFERENCES

Burks, R.L., E. Jeppesen, and D.M. Lodge. 2001. Pelagic prey and benthic predators: impact of odonate predation on *Daphnia*. *J. N. Am. Benthol. Soc.* 20(4):615-628.

Driver, E.A. 1977. Chironomid communities in small prairie ponds: some characteristics and controls. *Freshwater Biology* 7:121-133.

Edwards, T.D. and B.C. Cowell. 1992. Population dynamics and secondary production of *Hyalella azteca* (Amphipoda) in *Typha* stands of a subtropical Florida lake. *J. N. Am. Benthol. Soc.* 11(1):69-79.

Eiswirth, M., R. Ohlenbusch, and K. Schnell. 1999. Impact of chemical grout injection on urban groundwater. Impacts of Urban Grouth on Surface Water and Groundwater Quality (Proceedings of IUGG99 Symposium HS5, Birmingham, July 1999). IAHS Publ. no. 259.

Findlay, C.S. and J. Bourdages. 2000. Response time of wetland biodiversity to road construction on adjacent lands. *Conservation Biology* 14(1):86-94.

Hanes, T. and L. Stromberg. 1998. Hydrology of vernal pools on non-volcanic soils in the Sacrament valley. Pages 38-49 in: C.W. Witham, E.T. Bauder, D.Belk, W.R. Ferren Jr., and R. Ornduff (Editors). Ecology, Conservation and Management of Vernal Pool Ecosystems-Proceedings from a 1996 Conference. California Native Plant Society, Sacramento, CA.

Lillie, R.A. 2003. Macroinvertebrate community structure as a predictor of water duration in Wisconsin wetlands. *Journal of the American Water Resources Association* 39(2):389-400.

Mayes, W.M., A.R.G. Large, and P.L. Younger. 2005. The impact of pumped water from a de-watered Magnesian limestone quarry on an adjacent wetland: Thrislington, County Durham, UK. *Environmental Pollution* 138(3):443-454.

Meerhoff, M., C. Fosalba, C. Bruzzone, N. Mazzeo, W. Noordoven, and E. Jeppesen. 2006. An experimental study of habitat choice by *Daphnia*: plants signal danger more than refuge in subtropical lakes. *Freshwater Biology* 51:1320-1330.

Moorhead, D.L., D.L. Hall, and M.R. Willig. 1998. Succession of macroinvertebrates in playas of the southern high plains, USA. *J. N. Am. Benthol. Soc.* 17(4):430-442.

Murarka, I.P., T.E. Bailey, and J.R. Meiers. 2002. Water quality at a coal ash filled surface coal mine pit in Indiana. Proceedings of Coal Combustion By-Products and Western Coal Mines: A Technical Interactive Forum, April 16-18. Golden, CO.

Roadcap, G.S., W.R. Kelly, and C.M. Bethke. 2005. Geochemistry of extremely alkaline (pH>12) ground water in slag-fill aquifers. *Ground Water* 43(6):806-816.

Sutter, G. and R. Francisco. 1998. Vernal pool creation in the Sacramento valley: a review of the issues surrounding its role as a conservation tool. Pages 190-194 in: C.W. Witham, E.T. Bauder, D.Belk, W.R. Ferren Jr., and R. Ornduff (Editors). Ecology, Conservation and Management of Vernal Pool Ecosystems-Proceedings from a 1996 Conference. California Native Plant Society, Sacramento, CA.

USEPA. 1986. Quality Criteria for Water. Office of Water Regulations and Standards. EPA 440/5-86-001. Washington. DC 20460.

USEPA. 2006. National Recommended Water Quality Criteria. Office of Water. Office of Science and Technology (4304T).

Williams, D.D. 2005. Temporary forest pools: can we see the water for the trees? *Wetlands Ecology and Management* 13:213-233.

Yee, K.A., E.E. Prepas, P.A. Chambers, J.M. Culp, and G. Scrimgeour. 2000. Impact of $Ca(OH)_2$ treatment on macroinvertebrate communities in eutrophic hardwater lakes in the Boreal Plain region of Alberta: in situ and laboratory experiments. *Can. J. Fish. Aquat. Sci.* 57:125-136.

Zalom, F.G., A.A. Grigarick, and M.O. Way. 1980. Habits and relative population densities of some hydrophilids in California rice fields. *Hydrobiologia* 75:195-200.

Reviewed by Shawn Oliver (Bureau of Reclamation, Central California Area Office - Folsom) and John Wilson (Bureau of Reclamation, Dam Safety-Denver).

In: Dams: Impacts, Stability and Design
Editors: Walter P. Hayes and Michael C. Barnes

ISBN 978-1-60692-618-5
© 2009 Nova Science Publishers, Inc.

Chapter 3

BANKFULL HYDRAULIC GEOMETRY ADJUSTMENTS CAUSED BY CHECK DAMS FOR EPHEMERAL CHANNELS (SOUTH-EAST SPAIN, WESTERN MEDITERRANEAN)

C. Conesa-Garcia[1] and R. García-Lorenzo

Dept. de Geografía, Laboratorio de Geomorfología,
Universidad de Murcia, Murcia, Spain

ABSTRACT

Bankfull hydraulic geometry relationships, also called regional curves, relate bankfull channel dimensions to watershed drainage area. Nevertheless, construction of structures crossing the stream, such as check dams, can alter these relationships. This paper describes the bankfull hydraulic geometry adjustments caused by check dams for ephemeral channels in South-Eastern Spain, in particular for two of its torrential catchments most modified by projects of hydrological-forest restoration: i) the Cárcavo catchment, predominantly marly and lime-marly, and ii) the Torrecilla catchment, composed of metamorphic materials. The variations of hydraulic geometry at the bankfull stage between the reaches upstream and downstream from the check dams are analyzed in relation to the characteristics and spacing of the check dams, and especially in relation to other geomorphological variables also affected. Longitudinal bed slope, bed stability indexes, textural changes of deposits, dimensions of scour holes developed downstream from the check dams, upstream sedimentary wedges and refill degree of sediments behind these cross structures have been taken into account. Different regression equations were developed for both torrential streams with the aim of better defining how check dams affect ephemeral channels. Interesting results have been obtained from the bankfull hydraulic geometry relationships versus the watershed morphometry, the individual distance between check dams and the accumulated distances by consecutive refilled dams providing a continuous flow and sediment supply.

---

[1] corresponding author e-mail : cconesa@um.es.

Keywords: *Bankfull, Hydraulic geometry adjustments, Check dams, Ephemeral channels, South-East Spain*

INTRODUCTION

Check dams are structures of rock, wood, earth and other materials placed across the channel and anchored in the stream-banks to provide a "hard point" in the streambed that resists downcutting. Check dams can also reduce the upstream energy slope to prevent bed scour. Over the last few decades, these hydraulic structures have provided effective action in intermittent or ephemeral gullies that are actively downcutting (Castillo *et al.*, 2001; Conesa García, 2004). Rock sizes, spillways and crest designs for 25 -to 100- year flood events (FISRWG, 1998) are usually adopted, but this return period is often insufficient for torrential streams in arid and semi-arid environments.

Stream channel hydraulic geometry theory developed by Leopold and Maddock (1953) describes the interrelations between dependent variables such as width, depth and area as functions of independent variables such as watershed area or discharge. These relationships can be developed at a single cross-section (at-a-station) or across many stations along a reach (Merigliano, 1997). Hydraulic geometry relationships are empirically derived and can be developed for a specific river or watershed in the same physiographic region with similar rainfall/runoff relationships (Williams, 1978; Harman *et al.*, 1999).

Hydraulic geometry relationships are often used to predict channel morphology features and their corresponding dimensions. Bankfull hydraulic geometry relationships, also called regional curves, were first developed by Dunne and Leopold (1978) and related bankfull channel dimensions to drainage area (Andrews, 1980). Since then, studies on this subject, most of an applied nature, have proliferated (Merigliano, 1997; Harman *et al.*, 1999; Kuck, 2000; Johnson and Brown, 2001; Sweet and Geratz, 2003). The FISRWG (1998) suggests check dams as a priority objective for storm water management to reduce the frequency of bankfull flows in the contributing watershed. This is often done by constructing upstream storm water retrofit ponds that capture and retain increased storm water runoff for up to 24 hours before release. Gauge station analyses throughout the United States has shown that bankfull discharge has an average return interval of 1.5 years or 66.7% annual exceedance probability (Leopold, 1994). However, in ephemeral streams of arid and semiarid regime such return intervals can be much higher, reaching as much as 6 or 7 years in South-East Spain. However, using an existing check dam-influenced stream to develop basic data on bankfull area, bankfull depth, and other parameters useful for natural channel design can be challenging. The installation of check dams in a torrential watershed can result in important changes in hydraulic parameters, sediment load and gradation and scouring processes, which can in turn cause changes in stream morphology. The majority of the ephemeral streams subjected to hydrological correction and erosion control works are suffering an adjustment stage in response to such changes along the channel. As a result, streams affected by check dams can be unstable, and bankfull cross-section may be in the process of changing. In the case of the Torrecilla and Cárcavo *ramblas*, the Forest Hydrological Restoration Projects were mainly developed in the 70's.

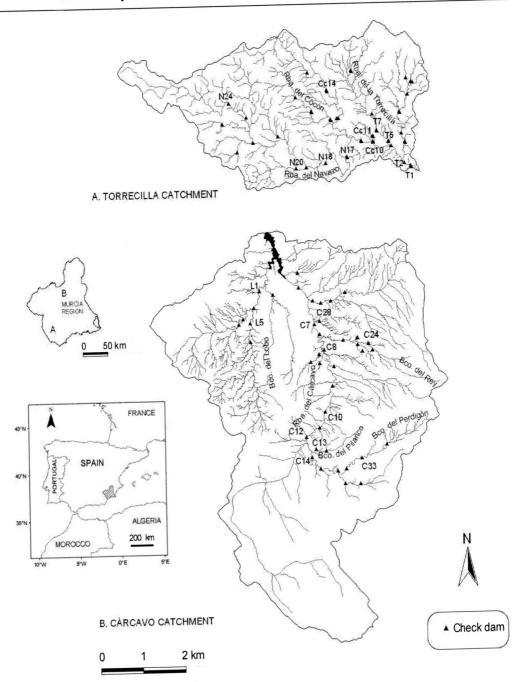

Figure 1. Location of the selected reaches in the Torrecilla and Cárcavo catchments. Triangular symbols represent the location of the check dams.

STUDY AREA

The catchments of the Torrecilla and Cárcavo *ramblas* are found in the Southeast of Spain and have a semi arid climate with a tendency to aridification. The mean annual precipitation is between 260 and 275 mm, annual ETP values are higher than 850 mm, with torrential precipitations occurring especially in the autumn and at the end of the summer. In this season the pluviometric intensities can reach more than 100 to 200 mm in a few hours, producing values of average pluvial erosivity per storm from 2.050 to 2.350 J m^2 (Conesa García and Alvarez Rogel, 2003). Both are catchments with torrential hydraulic conditions in which the relief, lithologic components and land uses establish their main differences (López Bermúdez *et al.*, 1998).

The Torrecilla catchment has an area of 15.5 km^2 and is situated in the southeast of the Region of Murcia (figure 1), within the interior domain of the Betic Cordilleras. The geological characteristics show complex structures eroded by ablation and metamorphic materials (slates, phyllites, schists and quartzites) mixed locally in sandstones and conglomerates that are rich in argillite. A dense network of drainage has been developed on these terrains, with rounded interfluves and scantily deepened channels. The high degree of bifurcation of this network, the erodibility of the schist and shale terrains and the high mean slope gradient of the basin (32 %) gives it a strong erosive potential.

To combat the erosion in this area the Spanish government through the Ministry of Agriculture approved and checked a forestry-hydraulic restoration Project, which includes the repopulation of 1237 ha with pine (*Pinus halepensis*) and the construction of check dams. To date 510.8 ha have been repopulated and a total of 33 check dams have been constructed; the majority of these between 1972 and 1979, four between 1986 and 1988 and others after 1995. Currently, the Torrecilla drainage network (96.37 km) has a density of 1 dam per each 2.09 km of channel.

The Cárcavo catchment (34.9 km^2) is situated in the western central part of the province of Murcia (38°13' N; 1°31' W). It mainly consists of limestone and marls Tertiary in age and alluvial Quaternary deposits. Locally gypsum is present. Relief is mainly formed by limestone and dolomite ridges. The lower parts are occupied by extended pediment surfaces that have been developed on the Miocene marls and Quaternary deposits. These pediments are for a great part in use for agriculture, or have been subject to reforestation with *Pinus halepensis* (Castillo *et al.*, 2001; RECONDES, 2005). Ephemeral channels and gullies dissect the pediment surfaces, producing very steep slopes.

This area was subject to a Forest Hydrological Restoration Project that started in 1969 and its revision began in 1977. During these projects about 38 check dams were constructed over the 148.4 km of the drainage network of this catchment (Fig. 1), which means a density of 1 dam per each 3.9 km of channel. In 1988, a large dam was built at the outlet of the main channel for flood control purposes.

METHODOLOGY

Field Indicators of Bankfull Stage

The correct identification of bankfull stage in the field can be difficult and subjective (Williams, 1978; Knighton, 1984; Johnson and Heil, 1996; MDE, 2000), and numerous definitions and methods exist for its identification in the field (Wolman and Leopold, 1957; Nixon, 1959; Schumm, 1960; Kilpatrick and Barnes, 1964; and Williams, 1978). The identification of bankfull stage in some regions is especially difficult due to dense vegetation and channel modification (e.g. Harman *et al.*, 1999). It is generally accepted that bankfull stage corresponds with the discharge that fills a channel to the elevation of the active floodplain. The bankfull discharge is considered to be the channel forming agent that maintains channel dimension and transports the bulk of sediment over time. Field indicators include the back of point bars, significant breaks in slope, changes in vegetation, the highest scour line or the top of the bank (Leopold, 1994). Nonetheless, these bankfull stage indicators are often more elusive in ephemeral channels (Rosgen, 1994). The most consistent bankfull indicators for streams affected by check dams are the highest scour line and the breaks in bank slope. It is rarely the top of the bank, especially in the immediate downstream reaches of the check dams, where generally there is an important incision on the bed. The indicator known as the "back of the point bar" has been used in upper reaches where the riffle-pool sequence is better represented. In these sections the bankfull channel edge is typically less distinct along the inner (point-bar) sides of stream bends, and often it has had to be determined by finding the same land elevation as that of the opposite bankfull channel edge.

In on-site inspection in wide stream bottoms, the edge of the bankfull channel is easily identifiable along the outer (current-bearing) banks of stream bends as the "crest" or "break" where streambank slope changes from being a steep channel wall to being the active flood flat. The active floodplain is the best indicator of bankfull stage (NWMC, 2005), but in the two *ramblas* of the study it is often difficult to identify bankfull discharge along steeper and narrower channel reaches. In places where the bankfull channel edge is not readily apparent it has used changes on the grain size distribution of surface deposits or changes from water-tolerant to upland plant species as indicators (TAC, 2003).

Data Collection

Field data were collected during the period from February 2002 to July 2005. For the analysis of bankfull geometry, 44 sections between dams (22 in the Torrecilla basin and 22 in the Cárcavo one) have been selected. Half of them were situated upstream (Up) from check dams and the other half downstream (Dw), in intermediate areas between the scour holes developed downstream of check dams and the sedimentary wedges retained behind these structures.

In the Torrecilla catchment eight of the 22 study sections are affected by four dams completely full of sediments (T7, Cc11, Cc14 and N17), eight more are influenced by check dams with a refilling to about 50 per cent (T25, N18, N20 and N24) and the rest depend on the control of check dams with scarce infill, lower than 20 per cent of its capacity (T2, T5 and T10). In the case of the Cárcavo stream nearly all the selected check dams are full; only two (C10 and C13) are not, although their infill degree is quite high (over 70 per cent). For each reach, a longitudinal survey was completed over a stream length equal to at least 20 bankfull widths (Leopold, 1994) and some bed features have been analyzed after the flow events: heads of riffles and pools, maximum pool depth, scour holes. The morphological variables change in short distances along the stream channel, as controls such as geology and tributaries join to the adjustments derived from the check dams.

Cross-sections were located in straight or non-sinuous reaches with uniform depth, avoiding proximity of gully junctions and turbulent sectors. Each cross-section selected for measurement was judged to be representative or typical after inspecting the entire reach. Reach uniformity was tested using HEC-RAS in consecutive cross-sections. Channel width, maximum and mean depth, wetted perimeter, entrenchment ratio, and flow area at bankfull stage were measured in 3-4 locations within the survey reach and averaged to produce representative reach values.

Representative reach values of flow area and wetted perimeter were used, together with measured profile slope, granulometric data and vegetal cover, to estimate Manning's roughness coefficient and bankfull discharge for different reaches. Particle-size distribution was determined for each reach using the pebble-count method and volumetric sampling. Ratios, cross-section areas, maximum and mean bankfull depths, bankfull mean velocity and discharge was determined with the aid of "*The Reference Reach Spreadsheet*" developed by Mecklenburg (1999). Morphologic variables required to classify the stream using Rosgen's method (Rosgen, 1996) have also been obtained to classify each stream reach. To this end, temporary pins were installed in the left and right banks, looking downstream, and from them the bankfull width and depth, and prone flood area were measured.

Active channel top width was quite subjective. When measuring width, visual judgment is an important element. In most cases, it is not obvious to delineate the bankfull channel width in the field. Each cross section survey started at or beyond the top of the left bank. Moving left to right, morphological features were surveyed including top of bank, bankfull stage, lower bench or scour, edge of water and channel bottom (Harrelson *et al.*, 1994; U.S. Geological Survey, 1969).

Data Analysis

The characterization of the change on bankfull cross-section was addressed by selecting relatively undisturbed, and comparing the data obtained with "regional curves" relating bankfull depth, width, and discharge to watershed area (Dunne and Leopold, 1978). Once representative data for bankfull area, depth and width were obtained from the local "reference reaches" upstream and downstream of check dams, the initial channel gradient and cross-sections were compared to these data, finding important differences.

From the survey data, at-a-station bankfull hydraulic geometry was calculated. To obtain a bankfull discharge (Q_{bf}) estimate, at these ungauged watersheds, Manning's equation was used. Once Q_{bf} was calculated in different referential reaches HEC-RAS was applied to choose the Q that is able to adapt better to bankfull conditions, i.e. the maximum Q that, within the rank of estimated flows, does not pass throughout the reach the bankfull level previously defined by the field indicators.

A vectorial information layer on channels, cross-sections, flow path and enbankments (if they exist) has been obtained using Arc-View and a digital elevation model generated by triangulation (TIN). The main advantage of Arc-View, in this case, is its facility to create the information layers needed by HEC-RAS. Once imported these covertures introduced the elevations of the cross-sections measured in the field, to complement those generated automatically from the DEM, as well as the roughness values (Manning coefficient) and bed slope. HEC-RAS also allows the representative bankfull level of the chosen cross-sections to be determined.

Table 1. Parameters of hydraulic geometry used in the calculation of bankfull discharge upstream and downstream from check dams. Values obtained for different stream reaches and refill percentages of the dam (Torrecilla catchment)

% refill	Part	Location*	Value	A (m²)	P (m)	Rh (m)	S (m/m)	n	v (m/s)	Q (m³/s)	q (m³/s)
100	Lower-Middle	Up	mean	1.55	9.01	0.171	0.023	0.045	1.05	2.43	0.27
			st.dev.	0.57	0.40	0.064	0.003	0.005	0.34	1.56	0.17
		Dw	mean	1.48	5.54	0.267	0.026	0.048	1.38	3.04	0.55
			st.dev.	0.47	1.59	0.101	0.011	0.004	0.26	1.43	0.28
> 50	Lower	Up	mean	1.55	9.01	0.171	0.023	0.045	1.05	2.43	0.27
			st.dev.	0.57	0.40	0.064	0.003	0.005	0.34	1.56	0.17
		Dw	mean	1.48	5.54	0.267	0.026	0.048	1.38	3.04	0.55
			st.dev.	0.47	1.59	0.101	0.011	0.004	0.26	1.43	0.28
	Middle	Up	mean	2.07	11.75	0.176	0.015	0.038	1.01	3.12	0.27
			st.dev.	0.17	3.04	0.032	0.008	0.003	0.34	0.80	0.14
		Dw	mean	1.70	5.40	0.316	0.030	0.058	1.37	3.48	0.64
			st.dev.	0.09	1.31	0.097	0.006	0.007	0.30	0.94	0.35
	Upper	Up	mean	0.45	2.40	0.187	0.044	0.063	1.09	0.73	0.30
			st.dev.	0.03	0.28	0.010	0.004	0.001	0.01	0.04	0.02
		Dw	mean	0.45	2.70	0.165	0.047	0.066	0.98	0.65	0.24
			st.dev.	0.01	0.28	0.030	0.005	0.008	0.11	0.06	0.01
< 20	Lower	Up	mean	1.49	6.15	0.242	0.031	0.044	1.54	3.40	0.55
			st.dev.	0.44	2.49	0.029	0.008	0.004	0.30	1.49	0.13
		Dw	mean	0.57	3.37	0.170	0.025	0.054	0.90	0.77	0.23
			st.dev.	0.07	0.51	0.017	0.010	0.006	0.06	0.07	0.01

* Location of reaches upstream (Up) and downstrean (Dw) of the check dams. A_{bf} = bankfull area; P_{bf} = bankfull perimeter; R_{bf} = bankfull hydraulic radius; S = bed slope; n = coefficient of roughness; v = mean velocity of flow; Q_{bf} = bankfull discharge; q_{bf} = unit bankfull discharge (Q_{bf} / P_{bf}).

In all the reaches the spacing between every par of consecutive cross-sections is equal or lower than 5 m, and yet new sections have also needed to be interpolated. Nevertheless, the simulation generated with HEC-RAS from measured and interpolated cross-sections does not always generates a good adjustment of the bankfull stage with only one field indicator. It has led in certain cases to using two or more indicators.

Besides the discharge graphical ouputs with information about bankfull flow area, wetted perimeter, hydraulic radius, velocity, among others variables, are also generated by this program. Such parameters are of great usefulness in the study of the hydro-morphological dominant processes of every reach.

In order to understand the effects of check dams on the longitudinal channel gradient, a new index has been applied, the so-called *Index of structural influence on the bed slope*

(*SIBS*), that connects the height of the check dams and bed gradient reduction (Conesa García *et al.*, 2005). This index was correlated with the average bankfull depth observed in the scour holes in order to explain its degree of incidence in downcutting processes and in the dimensions of the erosion wedge.

Several regression equations have been developed for both streams (Torrecilla and Cárcavo) to define the response of these ephemeral channels to the location of check dams. Relations such as the cross-section area and bankfull discharge with the watershed area (Aw) or the accumulated effect of the watershed area with the average watershed slope (Aw · Sw) of each check dam can be useful to understand the role of watershed areas in the refill and the geomorphologic effectiveness of these structures. With the same aim, bankfull geometrical parameters observed downstream of the check dams have been associated with the absolute distances between the structures and the cumulative distances between unsilted structures. The latter include intermediate segments with completely full dams that do not interrupt the continuity of flow or sediment transport.

Finally, some of the bankfull parameters of the improved reaches have been compared with the pre-dam stream geometry. In some of these segments it has not been possible to directly obtain the bankfull hydraulic radius previous to the construction of the check dams.

RESULTS AND DISCUSSION

Tables 1 and 2 summarize field measurements, hydraulic geometry and bankfull discharge for the Torrecilla and Cárcavo streams. Table 1 shows average values and their respective standard deviations of hydraulic geometry parameters measured upstream and downstream of check dams in the Torrecilla catchment, considering their refill level. Such values are the average of the measures and ratios obtained for the different cross-sections that make up the reaches (bankfull width and depth, bankfull cross-section area, width to flood prone area, entrenchment ratio, mean bankfull depth, width, width/depth ratio, pool and riffle features, ratio of lowest bank height to maximum bankfull depth, estimated mean velocity at bankfull stage and estimated discharge at bankfull stage). Although they are average values, it has been found that the refill percentage involves certain differences in the channel bankfull geometry that result in appreciable variations of the flow hydraulic system. It is worth remembering that these values are the averages of several cross-sections located in intermediate segments between the scour holes and the sedimentary wedges developed from the check dams.

Generally, the width/depth bankfull ratio is clearly higher upstream than in downstream dams. In reaches with completely infilled dams, the average hydraulic radius observed upstream of the structure is 26 per cent lower than downstream, and the reduction percentage approaches 40 per cent when dams are about half infilled. Reaches with empty or scarcely infilled dams are the exception, as bankfull is lower downstream of the dams (R_{bf}(Dw) = 0.17 m versus R_{bf}(Up) = 0.24 m). In this case, the area, the bankfull perimeter, the bed gradient and the flow velocity are lower too, a fact that can be explained by the high water retention capacity of these type of check dams, which are quite high (6-7 m) and are constructed with hydraulic masonry. At the base of these check dams a small scour hole may be developed partly due to the emergence of a sub-superficial flow which, after circulating under the dam, slightly scours the alluvial bed. In the upstream reaches a confined channel previous to check dam construction is quite common, whereas downstream, despite the absence of sediments from the upper segment, the gradient reduces (from 0.031 to 0.025) and, in a higher degree, the bankfull discharge (from 3.4 to 0.77 m^3/s). This is due to geological controls and the scarce or null effect on the bed slope by the recently installed check dams in lower parts of the watershed.

In the remaining areas of the watershed, downstream reaches from the dams represent mo-derately entrenched streams (entrenchment ratio: 1.41-2.2) (type B), whereas those located upstream show ratios greater than 2.2, i.e. they are slightly entrenched streams according to Rosgen's classification (1994). In this group we can distinguish types C and D in the most distant cross-sections of the sedimentary wedge and the E type in the vicinity of the dam dump, this last associated with the formation of pipes that act as drains of the low waters behind the structure.

The most important effects upstream from the check dams were a decreased gradient, width/depth ratio increase, reduced mean bed particle diameter and minor flow competence. The longitudinal channel profile changed causing headward aggradation. In all the cases, the bankfull channel area is higher upstream from the check dams due to the increase of roughness due to a denser vegetable cover and the lower gradient imposed by the sedimentary wedge. In reaches affected by check dams completely full of sediments the stream profile adopts a step-shape whose hydraulic projection is the height of the structure. Immediately downstream there is a longitudinal hole that is prolonged until the bed reaches a new balance profile. This is not possible until sediment load is re-established downstream achieving a balance with the rectified slope. The fact that these reaches register a greater bankfull area downstream (mean A_{bf} = 1.77 m^2; st. dev. = 0.43) is partly due to the advanced development of the scour holes, associated not only with the erosive effect of the artificial step, but also with a reduction on sediment load which increases the flow competence (mean Q_{bf} = 3.65 m^3/s for sections Dw).

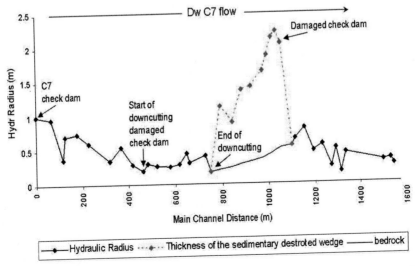

Figure 2. Variation of the bankfull hydraulic radius according to distance downstream from the check dam C7 (Lower Cárcavo).

Table 2. Hydraulic geometry parameters used to calculate bankfull discharge upstream and downstream from the check dams. Values obtained for different stream reaches (Cárcavo catchment)

Stream	Part	Location*	Value	A (m²)	P (m)	Rh (m)	S (m/m)	n	v (m/s)	Q (m³/s)	q (m³/s)
Main channel	Lower	Up	mean	5.20	15.70	0.331	0.009	0.061	0.74	5.75	0.37
		Up	st.dev.	4.02	14.90	0.270	0.012	0.055	0.83	4.97	0.33
		Dw	mean	3.06	8.17	0.374	0.022	0.058	1.31	5.97	0.73
		Dw	st.dev.	0.34	0.18	0.033	0.002	0.001	0.02	0.56	0.05
Main channel	Middle	Up	mean	2.42	10.22	0.237	0.023	0.055	1.04	3.77	0.37
		Up	st.dev.	0.19	1.02	0.005	0.002	0.001	0.09	0.03	0.04
		Dw	mean	1.92	5.90	0.326	0.028	0.057	1.38	3.93	0.67
		Dw	st.dev.	0.18	0.19	0.021	0.002	0.001	0.03	0.29	0.03
Left tributaries	Lower	Up	mean	1.63	9.47	0.172	0.065	0.053	1.50	3.63	0.38
		Up	st.dev.	0.41	0.18	0.047	0.054	0.013	0.06	0.67	0.08
		Dw	mean	1.58	7.29	0.217	0.048	0.050	1.57	3.71	0.51
		Dw	st.dev.	0.39	0.13	0.049	0.023	0.003	0.07	0.71	0.09
Right tributaries	Upper	Up	mean	0.59	3.91	0.151	0.034	0.055	0.95	0.83	0.21
		Up	st.dev.	0.40	2.75	0.027	0.005	0.003	0.14	0.56	0.07
		Dw	mean	0.52	2.86	0.184	0.042	0.057	1.16	0.91	0.32
		Dw	st.dev.	0.29	1.13	0.035	0.007	0.004	0.16	0.65	0.10

* Location of reaches upstream (Up) and downstrean (Dw) from the check dams.

In general, a similar pattern on geomorphic processes and hydraulic variables has been observed through the catchment, although some exceptions should be mentioned. For example, silty check dams (refill > 50 %) originate in the middle watershed reaches with a bigger bankfull area rather than in the lower part, which contrasts with the normal standard of sedimentary balance in natural streams. Until now, in the Torrecilla catchment the maximum stability level has not yet been reached, check dam filling is quite uneven depending on segments and tributaries, and the recent construction of bigger hydraulic structures in the lower streams increases the degree of general perturbation over the catchment. The proportion of silty dams in the middle reaches is higher than in the lower reaches. This explains why in the first case bankfull areas of 1.7 a 2.1 m² with scarce standard deviation (σ = 0.09-0.17) have been registered, whereas the lower streams hardly reach 1.5 m² of bankfull area (σ > 0.45).

The average values of bankfull discharge logically reflect such differences, but the variation, in both cases, between upstream and downstream cross-sections of the dams is also useful to confirm the fact that the average bankfull discharge is greater downstream than upstream of the structures. This leads us to suspect the presence of a sub-superficial flow fed by the water stored behind the check dam and favored by the higher permeability of the retained materials. This base flow emerges with relative frequency immediately downstream of the structure in alluvial beds with gradient higher than 0.01, and it seems to increase in direct relation to the volume of water retained and sediment texture, and inversely so with the bed slope. Therefore, the lower reaches present the greater additional amount of water all over the dam (0.61 m^3/s). They have a gradient between 0.01 and 0.026, accumulate detritus material in thick alluvial deposits and have enough storage capacity to feed infiltration. On the other hand, bankfull discharge generated downstream the empty check dams (refill < 20 %), located in the lower watershed part, are exclusively fed by this sub-superficial flow, which means 0.99 m^3/s.

In the Cárcavo catchment nearly all the check dams are infilled, so that the refill level ceases to be a distinguishing criterion of the geomorphologic influence in bankfull geometry. The characteristics of the structures and their location in the watershed context are therefore the main determinants of the variations in the channel bankfull geometry. Both the bankfull area and hydraulic radius and the discharge at bankfull stage are bigger in the lower main channel (A_{bf} = 3.06-5.20 m^2; R_{bf} = 0.3-0.4 m; 5.75-5.97 m3/s). Nevertheless, the destruction of a check dam can generate new bankfull adjustments due to regressive erosion processes, connecting with the scour wedge of the following upstream drop structure (Figure 2). Height diffe-rences of the check dams located along the main channel are not very evident and the catchment area influences the geomorphic and hydraulic bankfull features downstream. The sub-superficial discharge transferred from upstream of the check dam to downstream is lower to that estimated in the Torrecilla catchment: 0.16 and 0.25 m3/s in the middle and low reaches of the main channel respectively.

In all cases the bankfull area and wetted perimeter are greater upstream of the check dams. The least marked effects of the check dams in bankfull geometry have been observed in the headwater areas. This fact is partly deceitful, since it is due more to structural reasons (small check dams of gabions) and location than to the geomorphologic dynamics itself. For example, the check dams constructed in the upper right tributaries streams are, accor-ding to their minor height, too distant from each other. It attenuates their functionality in relation with the high transport capacity and bed instability along these reaches. Roughness coefficient, bankfull area and discharge are practically similar upstream and downstream ($n \approx$ 0,056; A_{bf} = 0.52-0.59 m^2 y Q_{bf} = 0.83-0.91 m3/s), whereas the bed slope and velocity are only lightly higher downstream of the check dams, contrary to the general tendency of the rest of the reaches.

The effectiveness of the check dams is also insufficient in the left tributaries of the Cárcavo stream. The effect, for example, in the channel bankfull dimensions is reduced along the lower Lobo gully, given the strong bed slope, scarcely corrected upstream, and the nature of the sediments, predominantly fine particles. In this reach, despite the fact that the bankfull area is lower than 1.7 m^2 the bankfull discharge reaches 3.6 m^3/s, with nearly no differences between the cross-sections situated upstream and downstream the check dams (ΔQ_{bf} = 0,08 m^3/s; $\Delta \sigma$ < 0,04).

Bankfull Hydraulic Geometry Relationships

The bankfull hydraulic geometry relationships represent 11 ungauged reaches for each watercourse, ranging in watershed area from 43 to 1088 ha in the Cárcavo catchment and from 4.5 to 827 ha in the Torrecilla catchment. Figure 3 shows hydraulic geometry curves for the Torrecilla catchment for mean bankfull discharge (Q_{bf}), unit bankfull discharge (q_{bf}) and bankfull cross-sectional area (A_{bf}) as a function of drainage area (A_w) and of product of this with the watershed slope ($A_w \cdot S_w$).

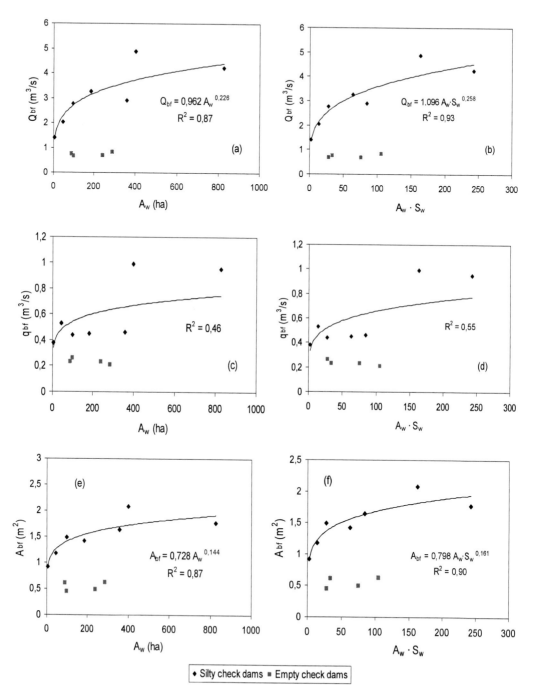

Figure 3. Torrecilla hydraulic geometry curves of mean bankfull discharge (Q_{bf}), bankfull unit discharge (q_{bf}) and bankfull cross-sectional area (A_{bf}) as a function of drainage area (A_w) and wathershed energy factor ($A_w \cdot S_w$).

The adjustment of the values has been done taking into account the reaches modified by silty check dams and eliminating the reaches affected by empty check dams. The relationship between bankfull discharge and drainage area is $Q_{bf} = 0.962\ A_w^{0.226}$ where Q_{bf} is bankfull discharge in cubic meters per second (m^3/s) and A_w is the watershed area in hectares. The regression equation has a determination coefficient (R^2) of 0.87. The fit for this equation type is improved if we relate the bankfull discharge and the product drainage area by watershed slope ($A_w \cdot S_w$). In this case, $Q_{bf} = 1.096\ A_w \cdot S_w^{0.258}$ and $R^2 = 0.93$. However, users must be careful to consider the variability represented by the 95% confidence limits for these relationships. Further work is necessary to develop reliable relationships from other managed semiarid catchments with higher amount of runoff data.

Bankfull cross-section area is the best geometric parameter in relation to the watershed area and the slope ($R^2 > 0.87$). On the other hand, unit bankfull discharges, which are estimated by wetted perimeter unit, show a very dispersed distribution in relation to these variables ($R^2 < 0.55$). Unit bankfull discharge seems to be very much controlled by geologic constrictions, modifying the normal development of the bankfull conditions. Bed narrows frequently take place caused by bedrock areas, which prevent the creation of flood active plains and favor the incision processes. Geo-structural narrows and bed armoring can be combined to produce anomalous bankfull geometry, represented by residual values very much separated from the trend curve, that relates it to the drainage area. The check dams influence channel geometry, because upstream they enlarge the channel through lateral erosion, and downstream they narrow it through scour processes. Therefore, the bankfull width is not the best variable to establish bankfull hydraulic geometry relationships in these corrected ephemeral channels.

The mean bankfull discharge of rectified reaches as function of area and hydraulic radius at bankfull stage for the Cárcavo and Torrecilla streams are shown in Figure 4. The best-fit regression equations belong to the relationships between bankfull area and discharge, which are represented by the distributions where the most of residual values are included between the upper and lower 95% confidence limits.

The power function regression equations and their coefficients of determination for the Torrecilla stream are:

$$Q_{bf} = 1.701\ A_{bf}^{1.379}\ ;\ (R^2 = 0.98) \tag{2}$$

$$Q_{bf} = 31.59\ R_{bf}^{2.048}\ ;\ (R^2 = 0.77) \tag{3}$$

And for the Cárcavo stream:

$$Q_{bf} = 1.74\ A_{bf}^{1.122}\ ;\ (R^2 = 0.95) \tag{4}$$

$$Q_{bf} = 69.63\ R_{bf}^{2.521}\ ;\ (R^2 = 0.90) \tag{5}$$

where, Q_{bf} = bankfull discharge (m^3/s), A_{bf} = bankfull cross-sectional area (m^2), and R_{bf} = bankfull hydraulic radius (m). Residual values are very dispersed in the relationships of the bankfull discharge versus hydraulic radius in the Torrecilla stream. This corroborates the previous expressed idea concerning the important influence of bedrock outcrops and check dams on channel width. These are constantly responding to new variations in the bankfull width, modifying also the values of hydraulic radius.

The natural variability resulting from variations in magnitude and frequency of runoff events, stream type and land uses is strongly affected by the degree of geomorphological effectiveness of the check dams. This fact is more clearly observed in the Torrecilla stream, where reaches with different phases of morphological adjustment, associated to unequal effects of the structures, coexist in its drainage network.

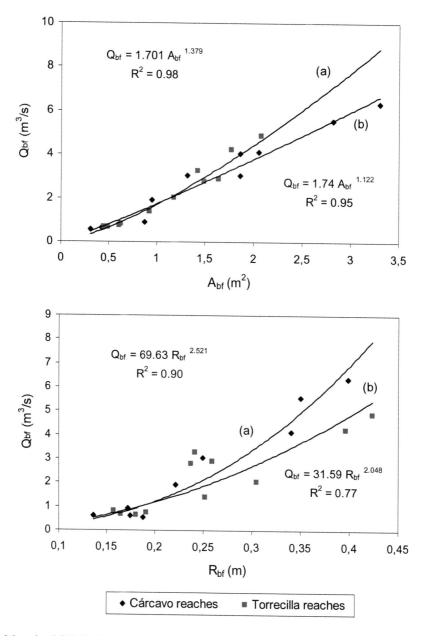

Figure 4. Mean bankfull discharge (Q_{bf}) of corrected reaches as function of area (A_{bf}) and hydraulic radius at bankfull stage (R_{bf}) in the Cárcavo and Torrecilla streams.

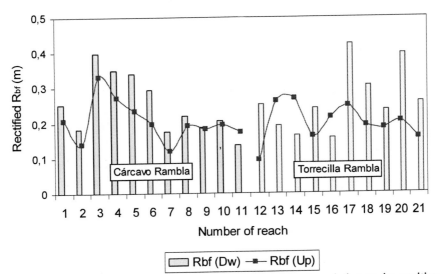

Figure 5. Average values of hydraulic radius bankfull for rectified reaches upstream and downstream of check dams in the Cárcavo and Torrecilla streams.

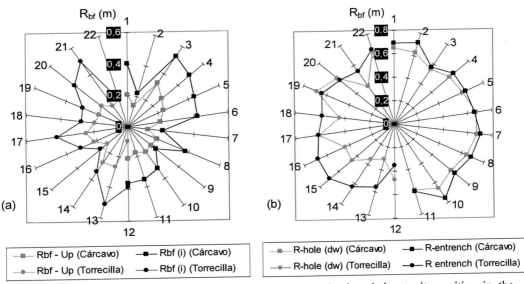

Figure 6. Bankfull hydraulic radius average values for study reaches before and after the construction of check dams in the Cárcavo and Torrecilla streams. R_{bf} (up) is the average hydraulic radius of each upstream corrected segment from the check dam; R_{hole} (dw) is the average hydraulic radius located just downstream from the check dam along the scour holes; and R_{bf} (i) is the initial bankfull average hydraulic radius.

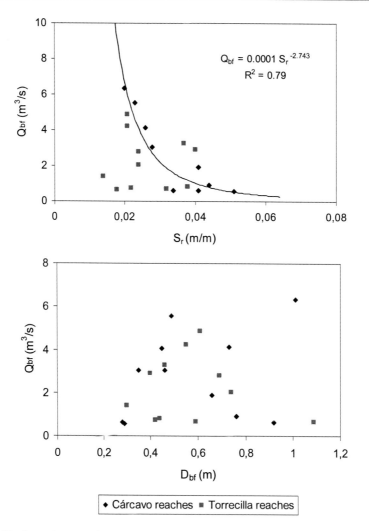

Figura 7. Relación de parámetros de geometría hidráulica con el caudal bankfull en las cuencas de la Torrecilla y del Cárcavo. Abf = el área de bankfull actual; Sr = la pendiente rectificada por la construcción de check dams; Dbf = la profundidad bankfull actual.

The slope of the regression curve for bankfull discharge (Q_{bf}) of the Torrecilla stream as a function of hydraulic radius (R_{bf}) is very similar to the curve slope of the Cárcavo stream (Figure 4). However, the intercept for the Torrecilla stream curve is slightly higher than the Cárcavo site. If the Torrecilla curves are compared with those from Cárcavo, we observe that they are generally higher than expected, indicating higher hydraulic radius and larger channel dimensions, despite the fact that they represent a small watershed area. This fact is clearly justified because the average gradient is very much higher in the Torrecilla catchment.

In addition, the average hydraulic radius from the selected reaches in the Cárcavo stream is a little smaller in the upstream segments of check dams than in those downstream (average $R_{bf} = 0.20$ m for upstream reaches and 0.25 m for those downstream) (Figure 5). However, in the segments where there is a modified slope in the Torrecilla stream, a different pattern of behavior between the lower watershed part and the upper one was observed. Except in the first segment, represented in Figure 5 by number 12, the lower reaches (segments from 13 to

16) present a low hydraulic radius bankfull downstream, since they are influenced by the new empty check dams. On the other hand, in the middle and upper reaches (segments from 17 to 21), the opposite occurs, because the hydraulic radius is usually lower upstream of the structures (average R_{bf} = 0.18). This last effect is typical of the morphological adjustments produced by check dams which are completely full of sediments or those which are in an advanced phase of silting.

Figure 6 shows the variation of the bankfull hydraulic radius of the corrected reaches in relation to initial reaches, made before the construction of check dams. Figure 6a shows at both Cárcavo and the Torrecilla catchments a lower bankfull hydraulic radius of the upstream dam reach in relation to ones existing during the pre-dam channel conditions. The average hydraulic radius, in the corrected channel reaches as well as in the initial ones, is seemed in both streams (Rbf (r) ≈ 0.2 m and Rbf (i) ≈ 0.6 m), although the differences ranged from 0.24 to 0.58 in both cases. Figure 6b compares the differences between the initial bankfull hydraulic radius of each reach and the average hydraulic radius of the scour holes developed downstream from the check dams. The figure shows that the current phase of excavation, which is just downstream from the check dams, has a very similar hydraulic radius in the Cárcavo stream, although it is a little bit lower than in the pre-dam construction phase (average Rhole = 0.64 m). This difference is very clear in some corrected reaches of the Torrecilla stream, mainly in the lower watershed part (Rbf(i) – Rhole > 0.24) and in others which are in an advanced phase of stabilization (numbers 18 and 21 of the Figure 6b).

For the other bankfull geometry relationships, some trend curves (Figure 7) have been delineated with a poor adjustment. Considering the bankfull discharge of both watercourses in relation to bankfull depth and rectified channel slope, it is verified that there is a poor relationship between these variables. Similarly, the bankfull depth shows a poor relationship in relation to the bankfull discharge in both basins. In this case, the reason is the importance of the bedrock reaches downstream from the check dams.

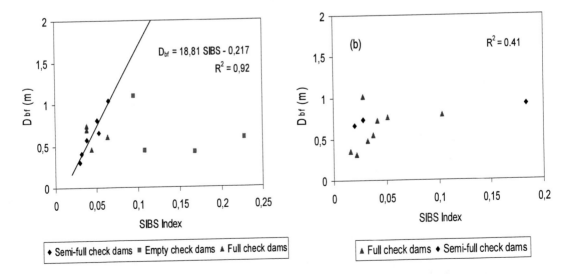

Figure 8. Bankfull channel depth of rectified reaches versus SIBS index en the Torrecilla (a) and Cárcavo (b) catchments.

In the Torrecilla streams, variations in the bed stability related to the different degree of geomorphological effectiveness of check dams are the main cause of the irregular distribution of the armored reaches. Bed armoring prevents the downcutting processes and stops the deepening of the channel, so that locally, in this case, check dams induce to lower bankfull depths than those naturally generated, so much upstream, through the effect of the sedimentary wedge, as downstream, through the flow scouring.

On the other hand, the rectified bed slope is related to the bankfull discharge just for the Cárcavo catchment ($R^2 = 0.79$). There, most of the check dams are full of sediments and the bed slope tends to be stabilized, reflecting its tendency in the area and present bankfull discharges. This does not occur in the Torrecilla catchment for the reasons previously mentioned. When we relate the bankfull channel depth and the SIBS index (*Structural influence on the bed slope*) in the downstream corrected reaches from the check dams, it is interesting to note that the best and in fact the only adjustment of the average values is registered in the segments affected by the semi-full check dams from the Torrecilla watercourse (Figure 8).

In this case, the relation between bankfull channel depth (D_{bf}) in meters and SIBS index (adimensionless) is expressed through a linear regression equation: $D_{bf} = 18.81$ SIBS $- 0.217$ ($R^2 = 0.92$). Each reach participates in the phase of maximum morphological alteration, since the quantity of sediments in suspension continues partially interrupted downstream from the dam, impeding the recovery of the bottom elevation in the scour holes and conserving its depth. In these conditions and only before the initiation of the channel stabilization phase, it is possible to observe the influence of the height of the check dams and, therefore, the reduction of the bed slope in relation to the variations of bankfull depth. From the 11 analyzed segments in the Cárcavo stream, just three are in this situation, which means it is impossible to establish a tendency.

Furthermore, the bankfull hydraulic geometry relationships have been studied as a function of the distances between the check dams and the accumulated distances in consecutive reaches whose check dams are totally silted and do not produce any interruption of the flow or the downstream sediment load. Table 3 shows the basic statistical parameters which are related to bankfull geometric variables in relation to the basin variables (A_w, $A_w \cdot S_w$) and the channel linear variables, particularly the distances between the check dams and the accumulated ones in those reaches where the check dams have lost their capacity to stop sediments.

In the Cárcavo basin the lower watershed part presents the greatest dimensions of the channel due to its greater drainage area. This does not occur in the Torrecilla basin, because the biggest average bankfull area and flow are in the middle watershed part. The reaches affected by check dams in the middle catchment have greater drainage areas than those which are located in the lower part, where the runoff has been partially fragmented. However, in both basins the lower reaches registered the most variable bankfull geometry values as a result of the different effect of the drop structures (st. dev. of $A_{bf} = 0.90$ for the Lower Cárcavo and 0.62 for the Lower Torrecilla). The accumulated watershed area (Ac. A_w) and the spacing between check dams also have very marked standard deviations in these cases; they can be higher than half of their corresponding average values (379 ha and 712 m respectively in the Cárcavo stream).

The mean bankfull geometry values obtained for the middle reach are very similar in both basins (A_{bf} between 1.63 and 1.96 and D_{bf} around 0.55). Moreover, their standard deviations are similar and something also the lowest of all the reaches. The main difference between both middle reaches is the distance of the accumulated thalweg in each check dam, it is to say the length of upstream flow path to the check dam. Check dams located upstream do not produce any interruption in this segment if they are silted. The accumulated average distance of the thalweg to the check dams in the middle part of the Carcavo stream (3238 m) is nearly the double of that which is in the same part of the Torrecilla stream (1675 m). Besides, the Carcavo middle reaches are characterized by a greater homogeneity in the functional channel length (st. dev. = 591 m) as well as in the accumulated watershed area (st. dev. = 91 ha). Similar values of the bankfull geometry for the upper watershed parts in both basins can be found, despite the environmental differences among them.

Table 3. Basic statistical parameters on bankfull hydraulic geometry in relation to watershed va-riables (A_w, $A_w·S_w$) and lineal variables of the corrected channel (distance between check dams)

	Part	Statis.	A_{bf} (m²)	D_{bf} (m)	R_{bf} (m)	Q_{bf} (m)	Ac. A_w (ha)	Distance (m)	Ac. dist. (m)	S_w (m/m)	$A_w·S_w$
Cárcavo catchment	Lower	media	2.32	0.60	0.30	4.74	717	1009	5217	0.146	98.8
		Dev-st	0.90	0.27	0.10	1.48	379	712	2023	0.031	47.2
		max	3.30	1.01	0.40	6.33	1088	2020	7393	0.189	140.0
		min	1.31	0.45	0.18	3.04	259	433	3109	0.121	49.0
	Middle	media	1.96	0.54	0.32	3.59	665	409	3238	0.122	85.2
		Dev-st	0.13	0.27	0.03	0.76	91	37	591	0.089	70.3
		max	2.05	0.73	0.34	4.13	729	435	3656	0.185	134.9
		min	1.86	0.35	0.30	3.05	601	383	2820	0.059	35.5
	Upper	media	0.59	0.58	0.19	0.92	156	375	1828	0.104	15.0
		Dev-st	0.29	0.29	0.03	0.56	138	59	698	0.030	10.8
		max	0.95	0.92	0.22	1.90	393	449	2501	0.150	32.6
		min	0.31	0.28	0.14	0.55	43	310	905	0.083	3.7
Torrecilla catchment	Lower	media	1.02	0.47	0.24	1.96	202	479	2261	0.393	74.2
		Dev-st	0.62	0.12	0.10	1.73	142	370	1460	0.080	56.3
		max	2.08	0.61	0.42	4.87	401	1200	3619	0.542	163.4
		min	0.49	0.30	0.16	0.68	4	166	166	0.315	2.4
	Middle	media	1.63	0.55	0.30	3.30	428	1265	1675	0.271	118.5
		Dev-st	0.14	0.15	0.09	0.80	370	630	985	0.030	111.8
		max	1.77	0.69	0.40	4.23	827	1795	2456	0.294	243.3
		min	1.49	0.40	0.24	2.78	98	569	569	0.237	27.6
	Upper	media	0.81	0.92	0.24	1.35	436	1467	1467	0.299	125.8
		Dev-st	0.51	0.25	0.09	0.98	553	820	820	0.017	158.0
		max	1.17	1.09	0.31	2.04	828	2046	2046	0.311	237.5
		min	0.45	0.74	0.18	0.66	45	887	887	0.287	14.0

Distance = spacing between check dams; Ac. dist. = accumulated distance or length of the functional channel for every check dam; Ac. A_w = accumulated watershed area in each check dam.

Bankfull hydraulic geometry relationships as a function of the distances between check dams and the accumulated distances from silty check dams upstream have been surveyed (Figure 9). In the Carcavo stream, the bankfull area and hydraulic radius in the downstream reaches from the check dams are closely related to the length of the accumulative thalweg, which depends on the total watershed area of each check dam. This relationship can be expressed by the following linear regression equations:

$$A_{bf} = 0.0005 \ ADI_{ChD} - 0.104; \ (R^2 = 0.90) \tag{6}$$

$$R_{bf} = 4E\text{-}0.5 \ ADI_{ChD} + 0.115; \ (R^2 = 0.81) \tag{7}$$

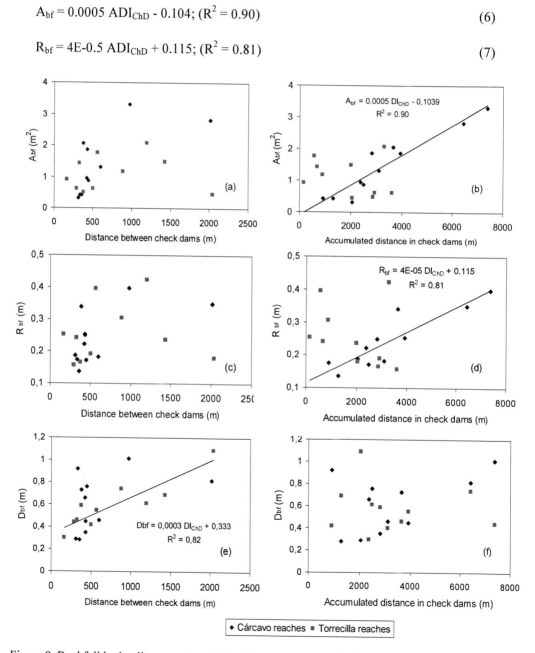

Figure 9. Bankfull hydraulic geometry relationships as a function of the distances between check dams and of the accumulated distances from silty check dams upstream.

A clear relationship of these variables does not exist in the Torrecilla stream and, however, it seems that the bankfull channel depth is better correlated to the specific distances between each pair of check dams. This may be explained by the presence of a higher number of non-silty check dams and, therefore, flow compartmentation, in addition to the morphological effects associated with an important sedimentary contribution of the own channel. In this case, the regression equation is $D_{bf} = 0.0003 \; DI_{ChD} + 0.333$ ($R^2 = 0.82$). The high coefficients of determination indicate a good agreement between the measured data and the best-fit relationships. However, the wide range of the values included within the 95% confidence limits indicates the need for caution when using these relationships.

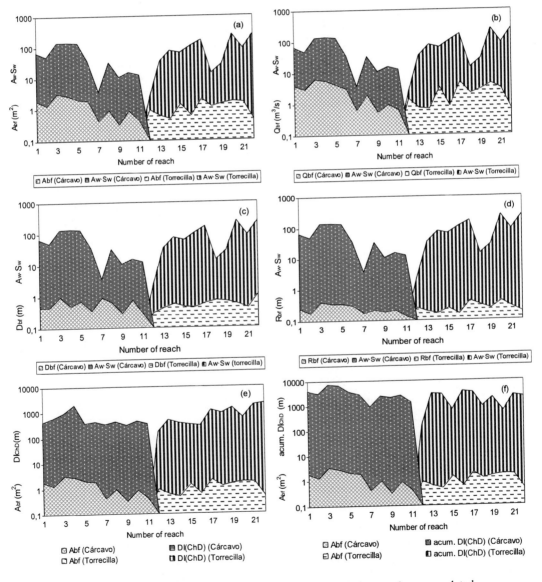

Figure 10. Profiles plotting the bankfull hydraulic geometry in comparison to the accumulated watershed area of check dams (ac.A_w), the watershed energy factor ($A_w \cdot S_w$). and the individual (DI_{ChD}) and accumulative distances between the check dams (acum. DI_{ChD}).

Figure 10 graphically illustrates the bankfull geometry variables in comparison to the accumulated watershed area of the dams (ac.A_w) and a topographic energy factor ($A_w \cdot S_w$). The enumeration of the ordinate axis (1-11 for the Carcavo stream and 12-21 for the Torrecilla stream) represents, in ascending order (from lower to upper reaches) the check dam position in each basin. Thanks to this procedure, it was checked if there are any comparable values between both kinds of variable (channel geometry and drainage area) and its spatial variation at the lower watershed part and the headwater sectors.

Figures 10a and 10b show similar profiles in the Carcavo catchment which represent the bankfull channel area downstream of check dams (A_{bf}) and the dimensional and topographic characteristics of the watershed area (A_w and $A_w \cdot S_w$). Moreover, it was observed the following: i) a small initial inflection which belongs to the lower reaches of the Barranco del Lobo, the main left tributary of the Carcavo stream, ii) a higher segment which include the lower reaches of the *rambla* main channel, iii) a new inflection with a peak referring to the middle reaches of the stream and the lower tributary gullies and, finally, a descent referring to the headwater areas. The correspondence is smaller in the case of the Torrecilla catchment, mainly in the lower part where the bankfull channel area shows a greater irregularity in relation to the space distribution of the drainage areas.

The space distribution profiles of the bankfull channel depth are not related to the extension and slope profiles of the drainage areas in any of the analyzed basins. Even clear differences between both basins are appreciated: profile with a form of saw-teeth in Carcavo and round form in Torrecilla (Figure 10c). On the other hand, the bankfull hydraulic radius decreases slowly towards the Carcavo upper part, reflecting almost imperceptibly the variations of $A_w \cdot S_w$ factor (Figure 10d). The comparison between the bankfull channel area and the individual and accumulative distances betwcen the check dams (figures 10e and 10f) is more noteworthy. While the profile which represents the distribution of the bankfull areas maintains a similarity with the spacing between check dams in the Torrecilla stream, in the case of the Carcavo stream the greatest similarity is found between the profiles created by the values A_{bf} and the functional channel length, not compartmented by silty check dams.

CONCLUSIONS

The results of this study indicate a good fit for regression equations for some hydraulic geometry relationships in the study ephemeral channels (*ramblas*). The upstream bankfull channel area from check dams seems to be the parameter which is best related to the extension and slope of the direct watershed area. The bankfull width is very variable depending on the lithologic controls and on the local influence of check dams. The same trend occurs with the bankfull depth, whose variations are closely related to bed armouring processes developed downstream of check dams. At most, a slight correlation of the bankfull width with the space between check dams in the Torrecilla stream was observed, perhaps because there is a great number of flow interruptions and sediment additions, and an important contribution of the global fluvial system to the interior channel modifications. On the other hand, the morphological channel adjustments produced upstream and downstream from check dams are significant in both *rambla* basins, but they show a great contrast in relation to their environmental features. The flow is torrential in both basins, but the nature and productivity of the sediment source-areas are different, as well as the slopes and land uses. In the Cárcavo stream, the transport of sediments in suspension predominates, while in the Torrecilla stream the bedload is high. In the first case, local erosion from the check dams produces larger scour holes than in the Torrecilla stream, where the thick material of the bed constitutes an significant obstacle to incision. Nonetheless, the average bankfull hydraulic radius measured outside the local alteration reaches is very much smaller than the initial one, previous to the construction of the dams. This demonstrates the effectiveness of check dams to control bed downcutting processes. Overall analysis of the results allows us to recognize the different phases of geomorphologic effectiveness in which the studied catchments are to be found. The Cárcavo stream is close to reaching the maximum stability level attributable to the hydrologic rectification works which exist in its catchment, but these are still insufficient to reach a suitable degree of stability. The Torrecilla stream presents some rectified channel reaches with uneven structural effects. It has relatively ineffective dikes in headwater areas which are highly stabilized due to the reforestation carried out since the 1970s. It also has check dams which are completely full in the lower and middle watershed parts, whose affected reaches are in an advanced phase of stabilization, and besides it has empty check dams located in the lower reaches which are perhaps oversized and which can change the present equilibrium of the channel, creating a notable disturbance at short and medium term.

These results may contribute to a better understanding of bankfull characteristics in ungauged streams affected by check dams at South-east of Spain. Some of the obtained bankfull hydraulic geometry relationships will facilitate the interpretation and design of stream restoration projects, assessment of stream health and future project planning in corrected ephemeral channels with similar environmental conditions. Since all of the curves have been estimated using drainage area, watershed slope and spacing between check dams, it will be relatively simple to check the expected bankfull characteristics in this field.

AKNOWLEDGMENTS

This paper was carried out within the research project PI/13 framework. "Hydrological and geomorphological response to the fluvial torrential systems affected by forestry-hydrological restoration Works in semiarid catchments of south-eastern Spain", financed by the *Fundación Instituto Euromediterráneo de Hidrotecnia*, European Council, and *Comunidad Autónoma de la Región de Murcia*.

REFERENCES

Andrew, E. D. 1980. Effective and Bankfull Discharges of Streams in the Yampa River Basin, Colorado and Wyoming. *Journal of Hydrology* 46:311-330.

Castillo Sánchez, V., González Barberán, G., Mosch, W., Navarro Cano, J.A., Conesa García, C. y López Bermúdez, F. 2001. Seguimiento y evaluación de los trabajos de restauración hidrológivo forestal, en López Bermudez, F. (coord.): *Seguimiento y evaluación de los efectos sobre el medio natural de la sequía y los procesos erosivos en la Región de Murcia.*

Consejería de Agricultura, Agua y Medio Ambiente de la Región de Murcia, Murcia, cap. 3, pp. 167-233.

Conesa García, C. and Alvarez Rogel, Y. 2003. Energía y erosividad pluvial del otoño en la región de Murcia. Correlación con el índice de agresividad pluvial de Fournier. In: Guijarro Pastor, J.A.; Grimalt Jalavert, M., Laita Ruiz de Asúa, M., Alonso Oroza, S. (Eds), *El Agua y el Clima*. Publicaciones de la Asociación Española de Climatología (AEC), Serie A, núm. 3, pp. 177-188.

Conesa García, C., Belmonte Serrato, F. and García Lorenzo, R. 2004. Efectos de los diques de corrección hidrológico-forestal en la competencia y estabilidad de corrientes efímeras. Aplicación a la cuenca de la rambla de la Torrecilla (Murcia). In C. Conesa García and J.B. Martínez Guevara (Eds), *Territorio y Medio Ambiente: Métodos Cuantitativos y Técnicas de Información Geográfica*, Asociación de Geógrafos Españoles y Universidad de Murcia, Murcia, pp. 69-83.

Conesa García, C., López Bermúdez, F. and García Lorenzo, R. 2005. Bed stability variations after check dam construction in torrential channels (South-East Spain). *Sixth International Conference on Geomorphology*, September 7-11, 2005, Zaragoza.

Dunne, T., and L.B. Leopold. 1978. Water in Environmental Planning. W.H. Freeman Co. San Francisco, CA.

Federal Interagency Stream Restoration Working Group (FISRWG). 1998. *Stream Corridor Restoration: Principles, Processes, and Practices*. PB98-158348LUW. Federal Interagency Stream Restoration Working Group, Washington, DC.

Harman, W.A., Jennings, G.D. and Patterson, J.M. 1999. Bankfull Hydraulic Geometry Relationships for North Carolina Streams. *AWRA Wildland Hydrology Symposium Proceedings*. Edited By: D.S. Olsen and J.P. Potyondy. *AWRA Summer Symposium*. Bozeman, MT.

Johnson, P.A., and T.M. Heil. 1996. Uncertainty in Estimating Bankfull Conditions. Water Resources Bulletin. *Journal of the American Water Resources Association,* 32(6):1283-1292.

Johnson, P.A., and Brown, E.R., 2001. Incorporating uncertainty in the design of stream channel modifications. *Journal of the American Water Resources Association*, 37(5), 1225-1236.

Kilpatrick, F.A., and. Barnes, H.H. 1964. Channel Geometry of Piedmont Streams as Related to Frequency of Floods. Professional Paper 422-E. US Geological Survey, Washington, DC.

Knighton, D. 1984. Fluvial Forms and Process. Edward Arnold, London.

Kuck, T.D. 2000. Regional Hydraulic Geometry Curves of the South Umpqua Area in Southwestern Oregon, *STREAM NOTES* January 2000, Stream Systems Technology Center, Rocky. Mountain Research Station, Fort Collins, Colorado.

Harrelson, C.C., J.P. Potyondy, C.L. Rawlins. 1994. Stream Channel Reference Sites: An Illustrated Guide to Field Technique. *General Technical Report RM-245*. U.S. Department of Agriculture, Forest Service, Fort Collins, Colorado.

Leopold, L.B., and Maddock Jr. T., 1953. The Hydraulic Geometry of Stream Channels and Some Physiographic Implications. *U.S. Geological Survey Professional Paper 252*, 57 pp.

Leopold, L.B. 1994. *A View of the River*. Harvard University Press, Cambridge, Massachusetts, 298 p.

López Bermúdez, F., Conesa García, C. and Alonso Sarría, F. 1998. Ramblas y barrancos mediterráneos: medio natural y respuesta humana. *Mediterrâneo, Num.12/13: Desertificaçao*. Instituto Mediterrânico, Universidade Nova de Lisboa, pp. 223-242.

MDE (Maryland Department of the Environment). 2000. *Maryland's Waterway Construction Guidelines*. Maryland Department of the Environment, Water Management Division, Baltimore, MD

Mecklenburg, D. 1999. *The Reference Reach Spreadsheet. Software Version 2.1*. Ohio Department of Natural Resources, Columbus, Ohio.

Merigliano, M.F. 1997. Hydraulic Geometry and Stream Channel Behavior: An Uncertain Link. *Journal of the American Water Resources Association,* 33(6): 1327-1336.

National Water Management Center (NWMC). 2005. *Regional Hydraulic Geometry Curves NWMC Procedure*. Natural Resources Conservation Service. United States department of Agriculture. http://wmc.ar.nrcs.usda.gov/technical/HHSWR/Geomorphic/procedure.html

Nixon, M. 1959. A Study of Bankfull Discharges of Rivers in England and Wales. In *Proceedings of the Institution of Civil Engineers*, vol. 12, pp. 157-175.

RECONDES, 2005. *Conditions for Restoration and Mitigation of Desertified Areas Using Vegetation*. Specific Targeted Research or Innovation Project. Sustainable development, global change and ecosystems. Report March 2005, *Coord. J. Hooke*, University of Portsmouth, 92 p.

Rosgen, D.L., 1994. A Classification of Natural Rivers. *Catena*, 22(1994):169-199.

Rosgen, D.L. 1996. *Applied River Morphology*, Wildland Hydrology Books, Pagosa Springs, Colorado.

Schumm, S.A. 1960. The Shape of Alluvial Channels in Relation to Sediment Type. *U.S. Geological Survey*. Professional Paper 352-B. U.S. Geological Survey, Washigton, DC.

Sweet, W.V. and Geratz, J.W. 2003. Bankfull Hydraulic Geometry Relationships and Recurrence Intervals for North Carolina's Coastal Plain. *Journal of the American Water Resources Association*, Vol. 39, Num. 4, 861-871

Technical Advisory Committee (TAC). 2005. *Recommendations for Revisions in Stream Definitions and Field Verification Methods*. June, 2003. Report to the Town of Chapel Hill from the Technical Advisory Committee on the Town of Chapel Hill Land Use Management Ordinance, 18 p.

U.S. Geological Survey. 1969. *Techniques of Water-Resources Investigations of the United States Geological Survey: Discharge Measurements at Gaging Stations*. Book 3, Chapter A8. U.S. Geological Survey, Washigton, DC.

Williams, G.P., 1978. Bankfull Discharge of Rivers. *Water Resources Research* 14(6): 1141-1154.

Wolman, M.G. and Leopold, L.B. 1957. River Floodplains: Some Observations on their Formation. *USGS Professional Paper 282-C*. U.S. Geological Survey, Washington, DC.

In: Dams: Impacts, Stability and Design
Editors: Walter P. Hayes and Michael C. Barnes

ISBN 978-1-60692-618-5
© 2009 Nova Science Publishers, Inc.

Chapter 4

EMBANKMENT OVERFLOW PROTECTION SYSTEMS AND EARTH DAM SPILLWAYS

Hubert Chanson[1]

The University of Queensland, Brisbane QLD 4072, Australia

1. INTRODUCTION

1.1. Presentation

The storage of water is essential for providing our society with drinking and irrigation water reserves. Storage along a natural stream is possible if the hydrology of the catchment is suitable. Hydrological studies provide information on the water volumes and as well as on the maximum (peak) flow in the system. Often the stream runoff does not provide enough supply all year round, and an artificial water storage system (e.g. the reservoir behind a dam) must be developed. For design engineers, it is essential to predict accurately the behaviour of hydraulic structures under the design conditions, operation conditions and emergency situations. The design approach is based on a system approach. A hydraulic structure must be analysed as part of its surroundings and the hydrology of the catchment plays an important role. Structural and hydraulic constraints interact, and the design of hydraulic structures is a complex exercise altogether. For example, the construction of a dam across a river requires a hydrological study of a stream. If the catchment can provide enough water all the year around, the risks of exceptional, emergency floods must be assessed. The design of the dam is based upon structural, geotechnical and hydraulic considerations. Political issues may further affect the site location and the final decision to build the structure. A consequent cost of the dam structure is the spillway system designed to pass safely the maximum peak flood. In addition the impact of the dam structure on the upstream and downstream catchments must be considered : e.g., sediment trap, fish migration, downstream water quality, modifications of the water table and associated impacts (e.g. salinity).

---

[1] Email: h.chanson@uq.edu.au - Url: http://www.uq.edu.au/~e2hchans/ Ph.: + 61 7 3365 3516 - Fax: +61 7 4465 4599.

A common type of water storage structures is the embankment reservoir and levees. But, during the 19th century, numerous embankment dams failed in Europe and North-America (Table 1-1, Figure 1-1). The two most common causes of failures were dam overtopping and cracking in the earthfill. The former was often caused by inadequate spillway facility. The latter resulted from a combination of bad understanding of basic soil mechanics, poor construction standards, and piping at the connection between bottom outlet and earth material. In recent years, the design floods of a number of dams were re-evaluated and the revised flows were often larger than those used for the original designs. In many cases, the occurrence of the revised design floods would result in dam overtopping because of the insufficient storage and spillway capacity of the existing reservoir. Figure 1-2 illustrates the overtopping of the Glashütte embankment dam on 12 August 2002. Completed in 1953, the Glashütte dam was a small flood retention system located in the Elbe river, upstream of Dresden (Germany). The stepped spillway capacity became insufficient during a very heavy storm event in August 2002.

A related form of embankment is the "natural dams" formed by landslides and rockslides. For example, during the Chi-chi earthquake in Taiwan on 21 September 1999, the Chin-Shui and Ta-Chia rivers, and the Tzao-Ling valley were dammed by massive landslides (Hwang 1999). The Tzao-Ling valley was previously dammed by record landslides in 1943 and 1974. In Tajikistan, Lake Sarez was formed by a massive rockslide damming the Murgab river valley during a severe earthquake in 1911. The reservoir contains nearly 17 E+9 m^3 of water nowadays. These landslide dams might become overtopped. In August 1191, a natural dam formed across the Romanche river in France. The landslide dam failed during the night of the 14 to 15 September 1219. The city of Grenoble suffered a massive flood wave and several thousand people were killed by the flooding. More recently, the May 2008 earthquake in China' Sichuan Basin created several lakes which became hazards and some were artificially breached..

Figure 1-1. Ruptured Dale Dyke embankment dam (UK) looking upstream - The dam break took place shortly before midnight on 11 March 1864 - The failure was be attributed to poor construction standard and cracks in the embankment close to the culvert.

Table 1-1. Examples of embankment dam failures and overtopping

Dam (1)	Construction date (2)	Date of accident (3)	Description of failure (4)	Loss of life (5)
Blackbrook dam, UK	1795-1797	1799	Collapse caused by dam settlement and spillway inadequacy.	None
South Fork (Johnstown) dam, USA	1839	May 1889	Overtopping and break of earth dam caused by spillway inadequacy.	over 2,000
Bilberry dam, UK	1843	5 Feb. 1852	Failure of earth dam caused by poor construction quality.	81
Dale Dyke dam, UK	1863	11 March 1864	Earth embankment failure attributed to poor construction work. Surge wave volume ~ 0.9 Mm^3.	150
Habra dam, Algeria	1873	December 1881	Break of masonry gravity dam caused by inadequate spillway capacity leading to overturning. Note that the storm rainfall of 165 mm in one night lead to an estimated runoff of about three times the reservoir capacity.	209
Dolgarrog dams, UK	1911/1910s	1925	Sequential failure of two earth dams following undermining of the upper structure.	25
Belci dam, Romania	1958-1962	1991	Dam overtopping and breach (caused by a failure of gate mechanism).	97
Teton dam, USA	1976	5 June 1976	Dam failure caused by cracks and piping in the embankment near completion.	11
Tous dam, Spain	1977	1982	Dam break (following an overtopping; collapse caused by an electrical failure)	None
Lake Ha! Ha! dam, Canada	--	July 1996	Dam overtopping caused by extreme rainfalls (18-22 July) in the Saguenay region.	None
Zeyzoun (or Zayaoun) dam, Syria	1996	4 June 2002	Embankment dam cracks, releasing about 71 Mm^3 of water. A 3.3-m high wall of water rushed though the villages submerging over 80 km^2. The final breach was 80-m wide.	22
Glashütte dam, Germany	1953	12 Aug. 2002	Embankment dam overtopping during very large flood because of inadequate spillway capacity.	None

References : Smith (1971), Schnitter (1994), Chanson (2004a)

In recent years, a number of overtopping protection systems were developed for embankments and earthfill dams. These include concrete overtopping protection systems, timber cribs, sheet-piles, riprap and gabions, reinforced earth, Minimum Energy Loss weirs, embankment overflow stepped spillways and the precast concrete block protection systems developed by the Russian engineers. Figure 1-3 shows a related design of rough chutes installed along a storm waterway. After a brief discussion of the embankment overtopping

breach process, several overflow protection systems are presented and detailed in the next
paragraphs.

Figure 1-2. Glashütte embankment dam overtopping (Germany) on 12 August 2002, shortly before dam
failure (Courtesy of Dr Antje Bornschein) - Dam overtopping was caused by the inadequate discharge
capacity of the spillway channel (in foreground).

Figure 1-3. Storm waterway with rough chutes at Michigan Drive, Oxenford QLD (Australia), view
from downstream in April 2003.

1.2. Embankment Dam Overtopping and Breach

The overtopping of an embankment is a relatively slow process compared to the failure of a masonry dam. It is not comparable to a sudden, explosive failure. For example, the failure of the 100 m high Teton dam started around 11:00 am on 5 June 1976 and the reservoir was drained by the evening. At its peak, the flow was estimated to be 28,000 m^3/s. During the failure of the Zeyzoun dam, the breach opened up to 6 m width about 3 and half hours after the initial breach. In the township of Ziara, 2 km downstream of the dam, the water depth peaked at about 4 m and dropped down to 10 cm a few hours later. In the case of the Glashütte dam (Figure 1-2), the downstream slope of the dam was grass-lined. Witness reports indicated that the dam was overtopped at 12:45 pm and that the wall failed completely within 30 minutes between 4:10 and 4:40 pm. That is, about 4 hours after the overtopping start (Bornschein and Pohl 2003). This last example illustrates the protection offered by some appropriate grass lining (Collett 1975, Hewlett et al. 1987). Hewlett et al. (1987) tested a variety of embankment slope reinforcement. Their results indicated that grass waterways, reinforced with concrete blocks, could sustain velocities up to 4 to 8 m/s without damage for a 3 hour test. Several grass species can be used : e.g. bent, couch, fescue, kikuyu, meadow grass, perennial ryegrass. Collett (1975) advised however against continual operation for more than a few days.

A number of recent studies investigated the overtopping and breach of embankment structures (Coleman et al. 2002, Rozov 2003). Both experimental studies under controlled laboratory flow conditions, and further prototype observations, showed that the embankment breach starts with an initiation phase, followed by a rapid development of the breach, and then an enlargement of breach width once the breach invert reaches the channel bed rock.

Inlet Shape of Natural Breach

Observations of natural scour in embankment breach showed a challenging similarity with the minimum energy loss weir inlet design during the breach development (Mckay 1970, Visser et al. 1990, Gordon 1981). Detailed breach data of non-cohesive embankment were recently re-analysed (Chanson 2004b). A half-flow net of breach inlet is presented in Figure 1-4A, showing some equipotentials and two streamlines. The breach contour lines are shown only below the water line. The flow cross-section areas and free-surface widths were measured along each equipotential, and the cross-sectional averaged Froude number and total head were calculated. Results are shown in Figure 1-4B where the Froude number and dimensionless total head H/H_1 are plotted as functions of the dimensionless centreline location, where H_1 is the upstream total head above downstream channel elevation and L is the embankment base length.

The re-analysis showed that the flow in the embankment breach is transcritical : i.e., 0.5 < Fr < 0.8 (Figure 1-4B) and the total head remains constant throughout the breach inlet up to the throat. Head losses occur downstream of the throat when the flow expands and separation takes place at the lateral boundaries. Separation is always associated with form drag and significant head losses, and the assumption of one-dimensional flow becomes invalid (Apelt 1983, Chanson 1999). Further the breach inlet lengths, measured along the breach centreline between inlet lip and throat, is about L_{inlet}/B_{max} = 0.5 to 0.6, where B_{max} is the free-surface width at the upper lip. The result is close to the minimum inlet length recommended for MEL culvert design : *"the minimum satisfactory value of length to inlet width L_{inlet}/B_{max} is 0.5"*

(Apelt 1983, p. 91). For shorter inlet lengths, flow separation may be observed in the inlet associated with unsatisfactory performances and energy dissipation.

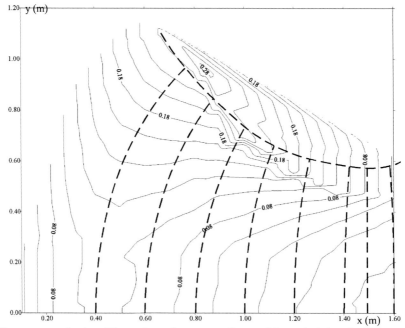

(A) Flow net analysis of breach and contour lines of breach inlet (half breach) - Flow from left to right - Equipotential lines: tick dashed lines, Bed contour elevations: thin solid lines - Note that contour lines above the free-surface are not shown.

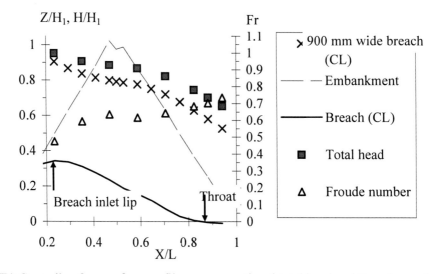

(B) Centreline free-surface profiles, cross-sectional total head and Froude number.

Figure 1-4. Inlet shape of a natural non-cohesive embankment breach during development (Data: Coleman et al. 2002) - 900 mm wide breach, t = 147 s, Q = 0.071 m^3/s, H_1 = 0.3 m, L = 1.7 m, coarse sand: d_{50} = 1.6 mm.

Breach Development Characteristics

During breach development, the outflow rate equals :

$$Q = C_D \times B_{max} \times \sqrt{g \times \left(\frac{2}{3} \times H_1\right)^3} \qquad (1.1)$$

where C_D is a dimensionless discharge coefficient ($C_D \sim 0.6$). During an overtopping event, the breach size increases with time resulting in the hydrograph of the breach. In Equation (1.1), the breach free-surface width B_{max} and upstream total head H_1 are both functions of time, embankment properties and reservoir size. For an infinitely long reservoir, a re-analysis of embankment breach data suggests that :

$$\frac{z_{lip}}{d_o} = 1.08 \times \exp\left(-0.0013 \times t \times \sqrt{\frac{g}{d_o}}\right) \qquad \text{for} \qquad t \times \sqrt{\frac{g}{d_o}} < 1750 \quad (1.2)$$

$$\frac{B_{max}}{d_o} = 2.73 \times 10^{-4} \times \left(t \times \sqrt{\frac{g}{d_o}}\right)^{1.4} \qquad \text{for } t \times \sqrt{\frac{g}{d_o}} < 1000 \qquad (1.3)$$

$$\frac{B_{min}}{d_o} = 4.01 \times 10^{-7} \times \left(t \times \sqrt{\frac{g}{d_o}}\right)^{2.3} \qquad \text{for } t \times \sqrt{\frac{g}{d_o}} < 1000 \qquad (1.4)$$

where d_o is the upstream flow depth, z_{lip} is the inlet lip elevation on the breach centreline and B_{min} is the free-surface width at the breach throat (Figure 1-4A). Note that Equations (1.2) to (1.4) were derived for cohesionless materials, and they are valid during the breach development only.

2. EMBANKMENT DAM OVERFLOW SYSTEMS (1) THE MINIMUM ENERGY LOSS WEIR DESIGN

2.1. Presentation

An unusual concept of overflow embankment spillway is the Minimum Energy Loss (MEL) weir design. The concept of the Minimum Energy Loss (MEL) structure was developed by late Professor Gordon McKAY (McKay 1971,1978) (App. I). The first MEL structure was the Redcliffe storm waterway system, also called Humpybong Creek drainage outfall, completed in 1960 in the Redcliffe peninsula (Australia). It consisted of a MEL weir acting as a streamlined drop inlet followed by a 137 m long culvert discharging into the Pacific Ocean. The weir was designed to prevent beach sand being washed in and choking the culvert, as well as to prevent salt intrusion in Humpybong Creek without afflux. The structure is still in use and passed floods greater than the design flow in several instances without flooding (McKay 1970, Chanson 2003,2007).

The concept of the Minimum Energy Loss (MEL) weir was developed to pass large floods with minimum energy loss and afflux ([2]), and nearly-constant total head along the waterway (Figure 2-1 to 2-3). The flow in the approach channel is contracted through a streamlined chute and the channel width is minimum at the chute toe, just before impinging into the downstream natural channel. The inlet and chute are streamlined to avoid significant form losses and the flow may be critical from the inlet lip to the chute toe at design flow. MEL weirs were designed specifically for situations where the river catchment is characterised by torrential rainfalls and by very small bed slope (Table 2-1). The first major MEL weir was the Clermont weir (Qld, Australia 1963) (Figure 2-2), if the small control weir at the entrance of Redcliffe culvert is not counted. The largest, Chinchilla weir (Qld, Australia 1973), is listed as a "large dam" by the International Commission on Large Dams (Icold 1984) (Figure 2-1). Figure 2-3 shows the ungated inlet at Lake Kurwongbah dam spillway. It was designed with the concept of minimum energy loss, in a fashion somehow similar to the design of a MEL culvert inlet (McKay 1971).

A MEL weir is typically curved in plan with converging chute sidewalls and the overflow spillway chute is relatively flat. The downstream energy dissipator is concentrated near the channel centreline away from the banks. At the chute toe, the inflow Froude number remains low and the rate of energy dissipation is small compared to a traditional weir. For example, the Chinchilla weir was designed to give no afflux at design flow (850 m³/s). In 1974, it passed 1,130 m³/s with a measured afflux of less than 100 mm (Turnbull and McKay 1974). Ideally a MEL weir could be designed to achieve critical flow conditions at any position along the chute and, hence, to prevent the occurrence of a hydraulic jump (Chanson 1999). This is not always achievable because the variations of the tailwater flow conditions with discharge are always important.

Figure 2-1. The Minimum Energy Loss weir on the Condamine river at Chinchilla (QLD, Australia 1973) on 8 Nov. 1997 during low overflow - Note the hydraulic jump in the foreground - Weir height: 14 m, Crest length: 410m, Spillway capacity: 850 m³/s.

---

[2] The afflux is the rise in upstream water level caused by the presence of the weir. Commonly used in culvert design, the afflux is a quantitative measure of the upstream flooding caused by the hydraulic structure.

(A) Model test for Q = 142 m³/s (prototype flow) (Collection of late Professor G.R. McKay, in Apelt (1978)).

(B) Weir overflow during a small to medium flood in 1993 (Courtesy of A.J. Holmes) - View from the right bank looking upstream.

Figure 2-2. Sandy Creek MEL weir, Clermont QLD (Australia).

Figure 2-3. Minimum Energy Loss spillway of Lake Kurwongbah dam on Sideling Creek dam, Petrie QLD (Australia) on 12 Sept. 1999 - Design flow: 710 m³/s, Reservoir capacity : 15.5 Mm³.

Table 2-1. Characteristics of Minimum Energy Loss weirs and spillways in Australia (all are still in use)

Structure	Q_{des} m³/s	H_{dam} m	B_{max} m	B_{min} m
(1)	(2)	(3)	(4)	(5)
MEL weirs				
Redcliffe QLD 1959	25.8	1.2	19.5	5.5
Sandy Creek weir, Clermont, Central Qld 1962-63	849.5	6.1	115.8	< 53 m
Chinchilla weir, Chinchilla QLD 1973	850.0	14.0	410.0	--
Lemontree weir, Condamine QLD 1979	--	4.0	--	--
MEL spillways				
Lake Kurwongbah, Sideling Creek dam, Petrie QLD 1958-69	849.5	25.0	106.7	30.5
Swanbank Power House, Ipswich QLD 1965	160.0	~ 6 to 8	45.7	7.31

Notes: Q_{des} : design discharge; B_{max} : crest width; B_{min} : chute toe width; H_{dam} : dam height above foundation.

2.2. Prototype Experiences

The first MEL structures were designed with the concept of constant total head, hence zero afflux, associated with solid physical modelling. MEL weir designs were scaled typically at 1:48 undistorted scale with fixed bed (Figure 2-2A). The MEL weirs are typically earthfill structures protected by concrete slabs and the construction costs must be minimum. The characteristics and operational record of a number of MEL structures were documented, and this was complemented by field inspections, surveys and oral discussions with designers (Chanson 2003).

Several structures were observed operating at design flows and for floods larger than design : e.g., Clermont and Chinchilla weirs. Inspections during and after flood events demonstrated a sound operation associated with little maintenance. Professor Colin Apelt stressed that improper inflow conditions could affect adversely the chute operation. Streamlining the inflow is essential. The successful operation of several structures for over 40 years has highlighted further considerations. MEL weirs are typically earthfill structures and the spillway section is protected by concrete slabs. An efficient drainage system must be installed underneath the chute slabs. Construction costs are minimum. A major inconvenient however is the overtopping risk during construction : e.g., Clermont weir in April 1963, Chinchilla weir twice in 1972 and 1973.

2.3. Design Considerations

The purpose of a MEL weir is to minimise afflux and energy dissipation at design flow conditions (i.e. bank full), and to avoid bank erosion at the weir foot. The weir is curved in plan to converge the chute flow and the overflow chute is relatively flat. Hence any downstream hydraulic jump is concentrated near the river centreline away from the banks and it is best located on, rather than downstream of, the chute toe. The inflow Froude number remains low and the rate of energy dissipation is very small compared to a traditional weir.

Assuming a broad-crest and no head loss at the intake (i.e. smooth approach), the discharge capacity of the weir equals :

$$Q = B_{max} \times \sqrt{g} \times \left(\frac{2}{3} \times \left(H_1 - \Delta z_{crest} \right) \right)^{3/2} \tag{2.1}$$

where H_1 is the upstream head, Δz_{crest} is the weir height above chute toe and B_{max} is the crest width (Figure 2-4).

Ideally, a M.E.L. weir could be designed to achieve critical flow conditions at any position along the chute and, hence, to prevent the occurrence of a hydraulic jump. Neglecting energy loss along the chute, the channel width B at any elevation Δz above the weir toe should be :

$$B_{min} = B_{max} \times \left(\frac{H_{des} - \Delta z_{crest}}{H_{des} - \Delta z} \right)^{3/2} \qquad \text{Ideal conditions (2.2)}$$

where H_{des} is the design upstream head above chute toe. This relationship (Equation (2.2)) is only achievable for the design flow and a suitable tailwater level. Note that the variations of the tailwater flow conditions with discharge are always important and a weak jump takes place at the chute toe. The jump occurs in an expanding channel and the downstream flow depth is fixed by the tailwater conditions downstream of the hydraulic jump (Figure 2-4).

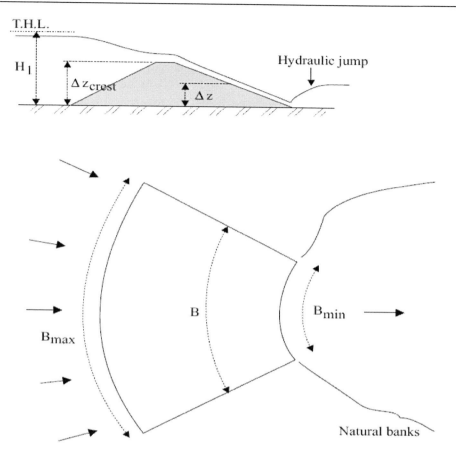

Figure 2-4. Definition sketch of a Minimum Energy Loss weir spillway.

In practice, the above pre-design calculations must be validated by physical model tests as illustrated in Figure 2-2A. Note further that the design may be extended for non-zero afflux design conditions, in a manner somehow similar to the design of MEL culverts (Chanson 2004c).

3. EMBANKMENT DAM OVERFLOW SYSTEMS (2) CONCRETE MACRO-ROUGHNESS ELEMENTS

Concrete macro-roughness linings provide a challenging alternative to embankment overflow spillway systems, with applications encompassing dykes, levees, road and rail embankments and dam spillways (Figure 1-3, 3-1 and 3-2). There are indeed numerous applications worldwide for both refurbishments and new structures. One such design is the baffle chute dissipator developed by the US Bureau of Reclamation (US Department of Interior 1965). Figure 3-1 presents a typical baffle chute dissipator down an embankment slope while Figure 3-2 shows a baffle chute down a steep catchment.

A macro-roughness lining system consists of precast concrete elements placed on a drainage-separation layer (US Department of Interior 1965, Montes 1998, Manso and Schleiss 2002). The stability concept is based on a combination of the self-weight of the blocks and of the interlocking anchors. Several types of elements may be used and are suitable for embankment slopes up to 1V:3H.

Design Considerations

In a baffle chute dissipator, the kinetic energy of the flow is dissipated by flow redirection over and around the baffle blocks arranged in offset rows. The design has no tailwater requirement, although a downstream transition section is required to protect the chute outlet from scour. The baffle chute design can pass most sediment and small to medium-size debris, but larger debris may become caught behind the blocks. Hence regular inspection and maintenance is necessary.

The concrete blocks must be adequately reinforced and tied with steel reinforcement to the chute floor. The block height h measured normal to the chute invert is typically larger than 80% of the critical depth at design flow. Each block is typically 1.5×h wide and the transverse distance between two adjacent blocks is also equal to 1.5×h except next to the wall. The chute walls are typically at least 3×h high, in the direction normal to the chute floor. Other dimensional requirements may be found in US Department of Interior (1965).

Manso and Schleiss (2002) tested a number of other block arrangements and proposed relevant design charts based upon physical model tests. Their study showed that the drainage layer lining is critically important for the stability of the concrete block chutes. The level of drainage had a significant impact on the uplift pressure forces on the foundation slabs and on the seepage flow within the embankment.

Figure 3-1. Baffle chute dissipator down an embankment slope in Irago peninsula (Japan) on 27 Mar.1999) - View from downstream with a small flow.

Figure 3-2. Baffle chute dissipator down Norman Creek upper catchment (Brisbane QLD, Australia) on 12 November 1996.

Practically, the macro-roughness elements are used for short chutes only (Figure 3-1 and 3-2). For longer chutes, the stepped overflow technique is cheaper to build and it can sustain larger discharges per unit width.

4. EMBANKMENT DAM OVERFLOW SYSTEMS (3) EMBANKMENT OVERFLOW STEPPED SPILLWAYS

4.1. Presentation

During the last three decades, a number of embankment dam spillways were designed with a concrete stepped spillway (Chanson 2001). During the 1990s, the construction of secondary stepped spillways accounted for nearly two-thirds of dam construction in USA (Ditchey and Campbell 2000). The preferred construction method was roller compacted concrete overlays placed on the downstream slope. Figures 4-1 to 4-3 show some examples. Figure 4-1 presents the Melton dam secondary spillway completed in 1994. With a discharge capacity $Q_{des} = 2,800$ m³/s, it is the world's largest embankment stepped spillway. Figures 4-2 and 4-3 present some recent structures designed as primary overflow stepped spillway.

There are several construction techniques to form a staircase embankment slopes, including gabions, reinforced earth, pre-cast concrete slabs and roller compacted concrete (RCC). RCC protection and gabion placement techniques yield embankment protections shaped in a simple stepped fashion. Gabion stepped chutes are usually restricted to small drops (Peyras et al. 1992) and the step face roughness must be accounted for (Gonzalez et al. 2008). Modern embankment dams are usually designed with a RCC stepped overtopping protection system. While most modern stepped spillways consists of flat horizontal steps, recent studies suggested different step configurations that might enhance the rate of energy dissipation (Andre et al. 2008, Gonzalez and Chanson 2008).

(A) General view of the secondary spillway on 30 January 2000 - Note the primary spillway in the background.

(B) Details of the steps (h = 0.6 m) next to the left abutment - Note the vertical step faces which were artificially roughened with the concrete formworks.

Figure 4-1. Overflow embankment dam stepped spillway: Melton dam (VIC, Australia) - Completed in 1916, the dam was heightened twice because of the rapid reservoir siltation - During the last refurbishment in 1994, the overflow stepped spillway was added.

Figure 4-2. Stepped spillway of Salado Creek Dam Site 15R (USA) (Courtesy of Craig Savela, USDA-NRCS-NDCSMC) - Roller compacted concrete construction, $\theta = 21.8°$, h = 0.61 m.

Figure 4-3. Construction of a stepped spillway for a detention basin in western Las Vegas (USA) (Courtesy of Mitchell R. Delcau, USACE Los Angeles District).

4.2. Construction Techniques

During construction, the concrete is placed in a succession of overlays of 0.2 to 0.4 m thickness and with a width greater than 2.5 m for proper hauling, spreading and compacting.

Recent projects favoured the use of roller compacted concrete. Roller compacted concrete (RCC) is defined as a no-slump consistency concrete that is placed in horizontal lifts and compacted by vibratory rollers. The primary advantages of RCC construction are the cost effectiveness and the short duration of construction works. The low cost of RCC works results from the lower cost per unit volume than conventional concrete and the construction technique with little formwork and placement with standard earthwork techniques.

For an embankment overtopping protection, exposed RCC is frequently used for secondary spillways with infrequent spills of less than 5 to 10 m^2/s. Experience of overflows over unprotected RCC faces were summarised by Chanson (2001). In cold climatic conditions, unprotected RCC surfaces can experience fracturing or chipping during freeze-and-thaw cycles. Some binders may be used to enable to entrain air into the roller compacted concrete and experience showed that 1.5 to 2% of air content is sufficient to protect concrete from frost deterioration. Alternatively, a conventional concrete protection overlay may be applied after the RCC or at the completion of construction works to protect the RCC.

With both RCC and conventional concrete protection, a drainage system beneath the concrete layers is essential to prevent uplift. Its purpose is to relieve pore pressure at the interface between the embankment and RCC overlays. In some cases, the drainage installation may be replaced or supplemented with drain holes formed through the RCC during placement. At the downstream end of the spillway, a cutoff wall must be built to prevent undermining of the concrete layer during overtopping.

Figures 4-2 and 4-3 show two RCC stepped spillways, while Figure 4-1 presents a conventional concrete structure. At the Melton dam (Figure 4-1), the cheapest tender proposed conventional concrete although the design specifications called for a RCC construction. It was suggested that the supplier preferred a stronger concrete resistance to allow access and passage for heavy equipment while the proximity of Melbourne CBD might have also influenced the low plain-concrete cost

4.3. Hydraulic Design

The embankment dam spillways are typically designed to operate in a skimming flow regime. During the design process, the embankment height, the downstream slope of the dam and the design discharge are generally given. The variable parameters may include the type of flow, the chute width and the step height. However, the designer is often limited to select a step height (h) within the values determined by the dam construction technique (h = 0.2 to 0.9 m with RCC). Gonzalez and Chanson (2007) detailed the complete design steps. A summary follows and a definition sketch is shown in Figure 4-4.

The location of the point of inception of free-surface aeration should be located upstream of the chute downstream end to ensure that the chute flow is fully-developed before the toe of the chute (Figure 4-4). Its characteristics may be calculated as:

$$\frac{L_I}{h \times \cos\theta} = 9.72 \times (\sin\theta)^{0.080} \times \left(\frac{q}{\sqrt{g \times \sin\theta \times (h \times \cos\theta)^3}} \right)^{0.71} \tag{4.1}$$

$$\frac{d_I}{h \times \cos\theta} = \frac{0.403}{(\sin\theta)^{0.04}} \times \left(\frac{q}{\sqrt{g \times \sin\theta \times (h \times \cos\theta)^3}}\right)^{0.59} \tag{4.2}$$

where q is the discharge per unit width (q = Q/B), L_I and d_I are the longitudinal distance from the crest and the flow depth at the inception point, g is the gravity acceleration and θ is the angle between the pseudo-bottom formed by the step edges and the horizontal.

If the channel is long enough for the flow to reach uniform equilibrium, the characteristic flow depth d equals:

$$d = \sqrt[3]{\frac{f_e \times q^2}{8 \times g \times \sin\theta}} \tag{4.3}$$

where f_e is the Darcy friction factor estimated based upon experimental air-water flow friction factor data as suggested by Chanson et al. (2002) and Chanson (2006). If the flow does not reach normal flow conditions before the toe of the chute, the flow depth must be deduced from the integration of the backwater equation :

$$S_f = -\frac{\partial H}{\partial x} = \sqrt{\frac{f_e}{8}} \times \frac{q^2}{g \times d^3} \tag{4.4}$$

Alternatively the flow properties in the gradually varied flow region may be calculated with a correlation curve linking some well-documented experimental results with the theoretical calculations in the developing and equilibrium flow regions (Gonzalez 2005) :

$$\frac{U_w}{V_{max}} = 0.00105 \times \left(\frac{H_1}{\sqrt[3]{q^2/g}}\right)^2 - 0.0634 \times \left(\frac{H_1}{\sqrt[3]{q^2/g}}\right) + 1.202 \tag{4.5}$$

where H_1 is the upstream total head above chute toe, d_c is the critical depth, V_{max} is the ideal flow velocity and U_w is the downstream velocity (Figure 4-4). In Equation (4.5), U_w is the unknown variable , H_1 and d_c are known, V_{max} is estimated from Equation (4.6). Once the dimensionless downstream velocity U_w/V_{max} is known, the flow properties can be estimated assuming fully developed flow conditions. The friction factor in skimming flow is typically f_e = 0.2. Finally, the ideal fluid flow velocity can be estimated from the Bernoulli equation:

$$V_{max} = \sqrt{2 \times g \times (H_1 - d \times \cos\theta)} \tag{4.6}$$

with d = q/U_w. This alternate method may be used for preliminary design calculations. It is important to note that it was obtained assuming f_e = 0.2 and it was only validated for skimming flow in stepped chutes with moderate slopes ($15° < \theta < 25°$).

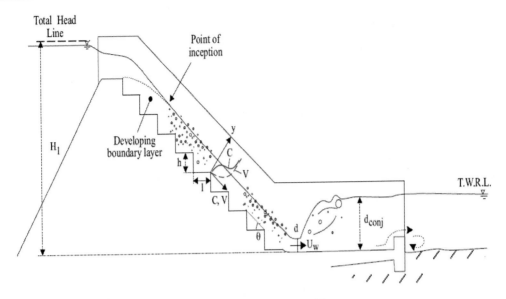

Figure 4-4. Definition sketch of an embankment dam stepped spillway.

Designers should be aware that the embankment overflow stepped spillway design is a critical process, as any failure can lead to a catastrophe. A number of key parameters should be assessed properly, including stepped face erosion, seepage through the embankment, drainage beneath the steps, interactions between the abutments and the stepped face ... In turn, some physical modelling with scaling ratios no greater than 3:1 is strongly advised.

Hydraulics of Small Embankment Dam Spillways

For short stepped chutes and large discharges, the flow may not be fully-developed before the downstream of the chute. That is, the chute length may be smaller than the distance between crest and inception point of free-surface aeration. A simple method was developed to predict the depth-averaged velocity and flow depth (Chanson 1999,2001).

In the developing flow region, the flow consists of a turbulent boundary layer next to the invert and an ideal-fluid flow region above. In the ideal-fluid region ($\delta < y < d$), the velocity, called the free-stream velocity, is deduced from the Bernoulli equation (Equation (4.6)). In the boundary layer, experimental data indicate that the velocity distribution follows closely a power law :

$$\frac{V}{V_{max}} = \left(\frac{y}{\delta}\right)^{1/N} \qquad\qquad 1 < y < \delta \ (4.7)$$

where y is the distance normal to the channel bed and δ is the boundary layer thickness. The velocity distribution exponent equals about $N = 5$ to 8 for stepped chutes. Combining Equations (6.6) and (6.7), the continuity equation gives:

$$q = V_{max} \times \left(d - \frac{\delta}{N+1} \right) \tag{4.8}$$

The boundary growth in skimming flow is enhanced by the turbulence generated by the steps. It may be estimated in first approximation :

$$\frac{\delta}{x} = 0.0301 \times \frac{1}{(\sin\theta)^{0.11}} \times \left(\frac{x}{h \times \cos\theta} \right)^{-0.17} \tag{4.9}$$

where x is the curvi-linear distance along the flow direction from the crest. Equation (4.9) was checked with model and prototype data (Chanson 1995, Meireles et al. 2006).

At a distance x from the crest, Equation (4.9) provides an estimate of the boundary layer thickness δ, and the flow depth d is given by Equation (4.8). The depth-averaged velocity is then $U_w = q/d$.

5. EMBANKMENT DAM OVERFLOW SYSTEMS (4) EARTH DAM SPILLWAYS WITH PRE-CAST CONCRETE BLOCKS

5.1. Presentation

A related form of concrete stepped overflow protection system is the earth dam spillway with pre-cast concrete blocks developed by the Russian engineers under the leadership of P.I. Gordienko (Gordienko 1978). The individual blocks are interlocked with the next elements in an overlapping staircase fashion and the stepped chute design assists in the energy dissipation (Pravdivets and Bramley 1989). For new dams, a stepped spillway made of concrete blocks may be considered as the primary flood release structure of the embankment (Figure 5-1). The design technique may be applied also to older embankment structures as at the Brushes Clough dam (UK) originally completed in 1859 and refurbished in 1991 with the construction of the new spillway on the downstream slope of the embankment (Figure 5-2). While the design concept was successfully used in Russia, and recently tested in USA and UK, it did not prove cost-effective in Europe nor North-America.

An interesting feature of the concrete block system is the flexibility of the stepped channel bed allowing differential settlements of the embankment. Individual blocks do not need to be connected to adjacent blocks. Another advantage is the fairly short construction time on site. The blocks are typically precast in factory and transported on site prior to installation.

Figure 5-1. Luhovitsy dam spillway in operation (Courtesy of Prof. Yuri Pravdivets).

Figure 5-2. Wedge concrete block spillway of the Brushes Clough dam, UK (1859/1991) in 1993 (Courtesy of Mr Gardiner, NWW) - The newer spillway is clearly visible on the downstream slope.

5.2. DESIGN AND CONSTRUCTION

For an embankment structure, the uppermost important criterion is the stability of the embankment material. Seepage may occur in saturated embankment and the resulting uplift pressures might damage or destroy the stepped channel and the whole structure. An adequate drainage is essential. In a typical design, the blocks lay on a filter and erosion protection layer. The layer has the functions to filter the seepage flow out of the subsoil and to protect the subsoil layer from erosion by flow in the drainage layer. Further the protection layer reduces or eliminates the uplift pressures acting on the concrete blocks. Usually a geotextile membrane is laid on the embankment before the placing of the layer, and another covers the protection layer before the installation of the blocks.

Suction of the fluid from underneath the concrete steps can be produced by the pressure differential created by the high velocity flow over the vertical face of the step. Drains placed in areas of sub-atmospheric pressure will function to relieve uplift pressures (Figure 5-3). In any case, the location of the embankment drains must be appropriately selected to avoid reverse flow in the drains and dynamic pressures associated with hydraulic jumps at low flows. Grinchuk et al. (1977) recommended that the total area of the drainage holes should be 10-15% of the exposed step area. More recently, Baker et al. (1994) suggested that an open area of 2 to 5% could be optimum. Frizell (1992) tested horizontal steps and sloped-downward steps (Δ = -10° and -15°). The model tests showed that the aspiration on the vertical step face increased with increasing downward slope of the steps.

The seepage flow in the embankment dam must be predicted accurately to make the appropriate provision for drainage and evacuation of seepage flow through the blocks. Note that the seepage flow may be influenced by the infiltration into the downstream slope caused by the spillway flow, in addition to the flow through the embankment.

Figure 5-3. Concrete block stepped spillway at the Dnieper test site (Courtesy of Prof. Yuri Pravdivets) - Wedge-shaped overlapping blocks (3 m × 3 m × 0.8 m) - Note the holes on the vertical step faces for drainage of the underlay.

Table 5-1. Design and stability of concrete blocks for embankment dam spillways

Reference	Maximum discharge (m2/s)	Slope degrees	Block geometry	Block thickness m	Comments
(1)	(2)	(5)	(4)	(5)	(6)
GRINCHUK et al. (1977)	up to 60 (V ≤ 23 m/s)	8.7	Wedge-shape reinforced concrete block.	0.7	Dneiper hydro plant. Horizontal steps.
PRAVDIVET S and BRAMLEY (1989)	2 to 27	7 to 15	Wedge-shaped block construction.	0.15 to 0.45	Inclined steps downward.
BAKER (1990)	--	21.8	Wedge-shaped blocks overlapping and non-overlapping ($\square$ = - 8.3°).		Model tests. No failure reported for : V ≤ 6.5 m/s.

Note : V : flow velocity.

Block Geometry

The Soviet engineers developed a strong expertise in the design of concrete wedge blocks and conducted extensive tests (Table 5-1). Pravdivets and Bramley (1989) described several configurations. The blocks are laid parallel to the slope on the top of the protective layer (Figure 5-4). Individual blocks do not need to be connected to adjacent blocks. The channel bed is very flexible allowing differential settlements. For large discharges, each block should be tied to adjacent blocks and made of reinforced concrete. Baker et al. (1994) suggested that a step height-to-length ratio in the range 1:4 to 1:6 ensures maximum stability of the blocks during overtopping.

Figure 5-4. Earth dam spillway made of precast concrete blocks : Zaraysk dam (also called Laraisky), Russia (Courtesy of Prof. Y. Pravdivets).

6. DESIGN CONSIDERATIONS

6.1 Alternatives for Embankment Overtopping

Alternative overtopping protection systems include timber cribs, sheet-piles, riprap and gabions, and reinforced earth (Chanson 2001). Timber crib overflows were used in Russia since the 18th century. Some recent structures are still in use (Chanson 2002). A number of embankment weirs were designed with steel sheet-piles and concrete slabs in Russia and Australia. An experimental structure was built with a reinforced-earth stepped overflow. A related design consists of pre-cast concrete panels set on very-coarse concrete acting as drainage layer between the embankment and the stepped overflow. Overflow system made with rockfill and gabions are another type of alternatives.

Figure 6-1 illustrates a few examples.

(A) Greenup timber crib weir (Inglewood QLD, Australia) in Sept. 1998.

(B) Stepped overflow at Robina QLD (Australia) in 1996 during construction.

(C) Neil Turner embankment weir near Mitchell QLD (Australia) (Courtesy of Chris Proctor).

(D) Guariaba gabion weir (Campo Grande, Brazil) (Courtesy of Officine Maccaferri).

Figure 6-1. Stepped overtopping protection systems for embankments.

6.2. Hydraulics Considerations

On a stepped spillway, the flow regime may be either nappe, transition or skimming flow depending upon the discharge per unit width and step geometry (Chanson 2001). Flow conditions in the transition flow regime exhibit some instabilities leading to deflecting nappes, fluctuating hydrodynamic loads on the steps, designers must avoid such conditions for the design flow rate and consider additional hydraulic and structural tests if they cannot avoid a transition flow regime at design conditions. Modern stepped spillways are typically designed for the skimming flow regime taking place for :

$$\frac{q}{\sqrt{g \times h^3}} > 1 \text{ to } 1.3 \tag{6.1}$$

where q is the discharge per unit width and h is the vertical step height

In the last five decades, most stepped spillways were designed with maximum discharge capacity less than 30 to 40 m²/s although there are few exceptions. Some ancient designs were designed and operated with larger discharges : e.g., the Mestalla weir (Spain, AD 960 with q up to 85 m²/s, the Quinson dam (France, 1870) with q up to 30 m²/s. More recently, prototype experiences with large overflows have included the Dneiper hydroplant in Russia (q ~ 60 m²/s, V = 23 m/s) (Figure 5-3) and the Shuidong and Dachaoshan dams in China. At the Shuidong dam, a major overflow took place for three days in 1994 and the spillway passed successfully discharges per unit width up to 90 m²/s without damage. In 2002, prototype tests were similarly conducted at the Dachaosham dam designed for q_{des} = 194 m²/s (LIN and HAN 2001)).

In skimming flows, the maximum acceptable flow rate is related to the step height and chute geometry. A detailed review of large roughness flows demonstrated that the flow resistance and the rate of energy dissipation decrease with decreasing relative roughness height (Chanson 1995, pp. 92-96). If the relative step height is small, the steps become a negligible roughness and no longer contribute to energy dissipation performances. Another limit is linked with the overall length of the spillway chute. Fully-developed flow conditions must be achieved before the downstream end of the chute to ensure adequate energy dissipation above the stepped structure (Chanson 2001). Both constraints yield a series of two equations in terms of the maximum acceptable flow rate down the stepped chute :

$$\frac{q}{\sqrt{g \times h^3}} < 60 \times (\cos\theta)^{1.5} \tag{6.2}$$

$$\left(\frac{q}{\sqrt{g \times h^3}}\right)^{2/3} < 0.119 \times \cos\theta \times (\sin\theta)^{0.26} \times \left(\frac{L}{h \times \cos\theta}\right)^{0.935} \tag{6.3}$$

where θ is the angle between the pseudo-invert formed by the step edges and the horizontal, and L is the spillway chute length measured between the spillway crest and stepped chute toe.

Equations (6.1) to (6.3) set the limits for the selection of the design discharge per unit width. If these conditions cannot be met, the chute design must be revised : e.g., by changing the vertical step height h, the channel width, the chute slope θ and the chute length L. Importantly, it must be understood that the largest discharge per unit width is not the most efficient flow situation in terms of energy dissipation.

6.3. Comments

With any embankment overtopping system, the stability of the earthfill/rockfill structure is essential including during overflow operation. Practical experiences highlighted the need for good construction quality together with simple, sound design. This is often associated with the selection of a straight prismatic chute operating with skimming flow. Known construction weaknesses include the drainage system, and the connection between the chute invert and the sidewalls. Further, at the downstream end of the chute, the residual energy of the flow must be dissipated safely in a stilling structure. This dissipation structure may be a hydraulic jump stilling basin (Figure 4-4) or a small flip bucket arrangement and a conventional concrete pool (Figure 5-2). Several prototype experiences including failures highlighted that the quality of the drainage layer is uppermost important (Pravdivets 1992, Baker et al. 1994, Chanson 2001).

The successful operation of the structure is also linked with matter-of-fact considerations. At an concrete block chute (Brushes Clough dam spillway), numerous acts of vandalism were reported, including destruction of concrete blocks. In parts of Africa and South-America, gabion structures were damaged by individuals using gabion wires and fillings for other purposes (e.g. fences, aggregates).

The interactions between seepage and overflow cannot be ignored in many applications, and practicing engineers do need some expert guidance. A number of researchers discussed the interactions between the seepage flow through the rockfill/earthfill structure and the overflow (e.g. Olivier 1967, Fenton 1968, Kells 1993). For example, Figure 6-1D illustrates some seepage on the downstream face of a gabion weir, with increasing seepage outflow toward the downstream weir toe. Recently, a 200 m long rockfill cascade was built for the diversion of Oaky Creek, Australia (Macintosh 2004). The physical tests showed significant interaction between seepage and overflow particularly for small to medium floods. With the stepped overflow design, these interactions are further complicated by the stepped flow patterns and significant rate of energy dissipation (Peyras et al. 1991,1992). Note that the influence of step face roughness on the overflow was specifically investigated by Gonzalez et al. (2008).

7. CONCLUSION

In recent years, the design floods of a number of dams were re-evaluated and the revised flows were often larger than those used for the original designs. In many cases, the occurrence of the revised design floods would result in dam overtopping because of the insufficient storage and spillway capacity of the existing reservoir. A number of overtopping protection systems were developed for embankments and earthfill dams, and they are discussed herein. These include concrete overtopping protection systems, Minimum Energy Loss weirs, embankment overflow stepped spillways and the precast concrete block protection systems.

For structures higher than 5 to 10 m, the stepped concrete overflow design is a sound technique that is well-suited to small to large discharges. The overflow cascading down the stepped spillway is characterised by some strong aeration, high turbulence of the flow and a significant rate of energy dissipation. A number of stepped spillways have been in operation for several decades although prototype experience of overflow remains relatively limited. Past failures and accidents suggested that a safe operation relies upon a sound design, a good quality of construction, suitable flow conditions and regular maintenance.

What is the best construction material ? There is no simple answer. For a temporary structure, a timber, timber crib or gabion spillway may be appropriate even for large overflows. For an embankment dams, unprotected roller compacted concrete (RCC) may be appropriate. Concrete stepped spillways are designed to last and they can pass large flow rates. For example, the Gold Creek dam spillway (1890) still in use, the Shuidong and Dachaoshan dam spillways operating with large flow rates. For aesthetical purposes, a cut-stone masonry construction can also provide good results.

APPENDIX I - PROFESSOR GORDON REINECKE MCKAY (1913-1989)

Born in Liverpool, Gordon Reinecke ("Mac") McKay was educated at Liverpool University in civil engineering, where he completed his Ph.D. in 1936. During his doctoral study, he visited Karlsruhe where he worked under the guidance of Professor Theodor Rehbock (1864-1950). In 1950, he moved to Australia where he became an academic staff of the NSW University of Technology, today the University of New South Wales, in Sydney. In 1951, he was appointed in the department of civil engineering at the University of Queensland (Brisbane) where he worked until his retirement in 1978. He was promoted to Professor in 1967.

Professor McKay contributed very significantly to the development of hydraulic physical models and design of hydraulic structures in Queensland. In the late 1950s and early 1960s, he developed the concepts of Minimum Energy Loss (MEL) culverts and weirs: i.e., Redcliffe MEL structure completed in 1960; Clermont weir completed in 1963. In 1980, the extension of the Hydraulics Laboratory at the University of Queensland (Brisbane QLD, Australia) was named the G.R. McKay Hydraulics Laboratory. In 1997, a creek in western Brisbane was named after Professor McKay : i.e., the McKay Brook in Kenmore.

REFERENCES

Andre, S., Bollaert, J.L., and Schleiss, A. (2008). "Ecoulements Aérés sur Evacuateurs en Marches d'Escalier Equipées de Macro-Rugosités - Partie 1: Caractéristiques Hydrauliques." Jl La Houille Blanche, No. 1, pp. 91-100.

Apelt, C.J. (1978). "A Commemorative Volume Presented to Professor Gordon R. McKay on the Occasion of his Retirement from the Department of Civil Engineering of The University of Queensland. 20th December 1978." Manuscript, Dept of Civil Engrg., Univ. of Queensland, Brisbane, Australia, 82 pages.

Apelt, C.J. (1983). "Hydraulics of Minimum Energy Culverts and Bridge Waterways." Australian Civil Engrg Trans., I.E.Aust., Vol. CE25, No. 2, pp. 89-95 (ISSN 0819-0259).

Baker, R. (1990). "Precast Concrete Blocks for High Velocity Flow Applications." J. IWEM, Vol. 4, Dec., pp. 552-557. Discussion : Vol. 4, pp. 557-558.

Baker, R., Pravdivets, Y., and Hewlett, H. (1994). "Design Considerations for the Use of Wedge-Shaped Precast Concrete Blocks for Dam Spillways." *Proc. Instn Civ. Engrs Wat. Marit. and Energy*, Vol. 106, Dec., pp. 317-323.

Bornschein, A., And Pohl, R. (2003). "Dam Break during the Flood in Saxon/Germay in August 2002." *Proc. 30th IAHR Biennial Congress*, Thessaloniki, Greece, J. Ganoulis and P. Prinos Ed., Vol. C2, pp. 229-236.

Chanson, H. (1995). "Hydraulic Design of Stepped Cascades, Channels, Weirs and Spillways." *Pergamon*, Oxford, UK, Jan., 292 pages (ISBN 0-08-041918-6).

Chanson, H. (1999). "The Hydraulics of Open Channel Flow : An Introduction." *Edward Arnold*, London, UK, 512 pages (ISBN 0 340 74067 1).

Chanson, H. (2001). "The Hydraulics of Stepped Chutes and Spillways." *Balkema*, Lisse, The Netherlands, 418 pages (ISBN 90 5809 352 2).

Chanson, H. (2002). "Timber Crib Weirs in Queensland, Australia. Some Heritage Stuctures with a Solid Operational Record." *Royal Historical Society of Queensland Journal*, Vol. 18, No. 3, pp. 115-129.

Chanson, H. (2003). "Minimum Energy Loss Structures in Australia : Historical Development and Experience." *Proc. 12th Nat. Eng. Heritage Conf.*, IEAust., Toowoomba Qld, Australia, N. Sheridan Ed., pp. 22-28 (ISBN 0-646-42775-X).

Chanson, H. (2004a). "Environmental Hydraulics of Open Channel Flows." *Elsevier Butterworth-Heinemann*, Oxford, UK, 483 pages (ISBN 978 0 7506 6165 2).

Chanson, H. (2004b). "Overtopping Breaching of Noncohesive Homogeneous Embankments. Discussion." *Jl of Hyd. Engrg.*, ASCE, Vol. 130, No. 4, pp. 371-374 (ISSN 0733-9429).

Chanson, H. (2004c). "The Hydraulics of Open Channel Flow : An Introduction." *Butterworth-Heinemann*, Oxford, UK, 2nd edition, 630 pages (ISBN 978 0 7506 5978 9).

Chanson, H. (2006). "Hydraulics of Skimming Flows on Stepped Chutes: the Effects of Inflow Conditions?" *Journal of Hydraulic Research*, IAHR, Vol. 44, No. 1, pp. 51-60.

Chanson, H. (2007). "Hydraulic Performances of Minimum Energy Loss Culverts in Australia." *Jl of Performances of Constructed Facilities*, ASCE, Vol. 21, No. 4, pp. 264-272 (DOI: 10.1061/(ASCE)0887-3828(2007)21:4(264)).

Chanson, H., and Toombes, L. (2002). "Experimental Investigations of Air Entrainment in Transition and Skimming Flows down a Stepped Chute." *Can. Jl of Civil Eng.*, Vol. 29, No. 1, pp. 145-156.

Chanson, H., Yasuda, Y., and Ohtsu, I. (2002). "Flow Resistance in Skimming Flows and its Modelling." *Can Jl of Civ. Eng.*, Vol. 29, No. 6, pp. 809-819.

Coleman, S.E., Andrews, D.P., Webby, M.G. (2002). "Overtopping Breaching of Noncohesive Homogeneous Embankments." *Jl of Hyd. Engrg.*, ASCE, Vol. 128, No. 9, pp. 829-838. Discusion: 2004, Vol. 130, No. 4, pp. 371-374. Closure: 2004, Vol. 130, No. 4, pp. 374-376.

Collett, K.O. (1975). "Unusual Surfaces for Large Spillways." *ANCOLD Bulletin*, No. 42, July, pp. 3-10.

Ditchey, E.J., and Campbell, D.B. (2000). "Roller Compacted Concrete and Stepped Spillways." *Intl Workshop on Hydraulics of Stepped Spillways*, Zürich, Switzerland, H.E. MINOR and W.H. HAGER Editors, Balkema Publ., pp. 171-178.

Fenton, J. D. (1968). "Hydraulic and Stability Analyses of Rockfill Dams." *Res. Rep.*, Univ. of Melbourne, Dept. of Civil Engineering, Melbourne, Australia.

Frizell, K.H. (1992). "Hydraulics of Stepped Spillways for RCC Dams and Dam Rehabilitations. " *Proc. 3rd Specialty Conf. on Roller Compacted Concrete*, ASCE, San Diego CA, USA, pp. 423-439.

Gonzalez, C.A. (2005). "An Experimental Study of Free-Surface Aeration on Embankment Stepped Chutes." *Ph.D. thesis*, Department of Civil Engineering, The University of Queensland, Brisbane, Australia, 240 pages.

GONZALEZ, C.A., and Chanson, H. (2007). "Hydraulic Design of Stepped Spillways and Downstream Energy Dissipators for Embankment Dams." *Dam Engineering*, Vol. 17, No. 4, pp. 223-244.

Gonzalez, C.A., and Chanson, H. (2008). "Turbulence and Cavity Recirculation in Air-Water Skimming Flows on a Stepped Spillway." *Jl of Hyd. Res.*, IAHR, Vol. 46, No. 1, pp. 65-72.

Gonzalez, C.A., Takahashi, M., and Chanson, H. (2008). "An Experimental Study of Effects of Step Roughness in Skimming Flows on Stepped Chutes." *Journal of Hydraulic Research*, IAHR, Vol. 46, No. Extra Issue 1, pp. 24-35.

Gordienko, P.I. (1978). "Reinforced-Concrete-Earth Overflow Dams." *Dams and Spillways*, Collection of Works No. 61, Issue 2, MISI, Moscow, pp. 3-17 (in Russian).

Gordon, A.D. (1981). "The Behaviour of Lagoon Inlets." *Proc. 5th Australian Conf. Coastal and Ocean Eng.*, Perth WA, pp. 62-63.

Grinchuk, A.S., Pravdivets, Y.P., and Shekhtman, N.V. (1977). "Test of Earth Slope Revetments Permitting Flow of Water at Large Specific Discharges." *Gidrotekhnicheskoe Stroitel'stvo*, No. 4, pp. 22-26 (in Russian). (Translated in Hydrotechnical Construction, 1978, Plenum Publ., pp. 367-373).

Hewlett, H.W.M., Boorman, L.A., and Bramley, M.E. (1987). "Design of Reinforced Grass Waterways." *CIRIA Report No. 116*, London, UK, 116 pages.

Hwang, H.Y. (1999). "Taiwan Chi-Chi Earthquake 9.21.99. Bird's Eye View of Cher-Lung-Pu Fault." *Flying Tiger Cultural Publ.*, Taipei, Taiwan, 150 pages.

International Commission on Large Dams (1984). "World Register of Dams - Registre Mondial des barrages - ICOLD." *ICOLD*, Paris, France, 753 pages.

Kells, J.A. (1993). "Spatially Varied Flow over Rockfill Embankments." *Can. Jl of Civ. Engrg.*, Vol. 20, pp. 820-827. Discussion : Vol. 21, No. 1, pp. 161-166.

Lin, Keji, and Han, Li (2001). "Stepped Spillway for Dachaoshan RCC Dam." *Shuili Xuebao* (Jl of Hydraulic Engrg.), Beijing, China, Special Issue IAHR Congress, No. 9, pp. 84-87 (in Chinese).

McKay, G.R. (1970). "Pavement Drainage." *Proc. 5th Aust. Road Res. Board Conf.*, Vol. 5, Part 4, pp. 305-326.

McKay, G.R. (1971). "Design of Minimum Energy Culverts." *Research Report*, Dept of Civil Eng., Univ. of Queensland, Brisbane, Australia, 29 pages and 7 plates.

McKay, G.R. (1978). "Design principles of Minimum Energy Waterways." *Proc. Workshop on Minimum Energy Design of Culvert and Bridge Waterways*, Australian Road Research Board, Melbourne, Australia, K.F. PORTER Ed., Session 1, pp. 1-39.

Macintosh, J. (2004). "Steep Gradient Waterway Stabilization - An Innovative Design Technique." *Proc. 8th National Conf. on Hydraulics in Water Engineering*, H. CHANSON and J. MACINTOSH Eds, IEAust., Gold Coast, Australia (CD-Rom).

Manso, P.A., and Schleiss, A.J. (2002). "Stability of Concrete Macro-Roughness Linings for Overflow Protection of Earth Embankment Dams." *Can Jl of Civil Eng.*, Vol. 29, No. 5, pp. 762-776.

Meireles, I., Cabrita, J., and Matos, J. (2006). "Non-Aerated Skimming Flow Properties on Stepped Chutes over Small Embankment Dams." *Proc. Intl Junior Researcher and Engineer Workshop on Hydraulic Structures*, IAHR, Montemor-o-Novo, Portugal, 2-4 Sept., J. MATOS and H. CHANSON Ed., Hydraulic Model Report No. CH61/06, Div. of Civil Engineering, The University of Queensland, Brisbane, Australia, pp. 91-99.

Montes, J.S. (1998). "Hydraulics of Open Channel Flow." *ASCE Press*, New-York, USA, 697 pages.

Olivier, H. (1967). "Through and Overflow Rockfill Dams - New Design Techniques." *Proc. Instn. Civil Eng.*, March, 36, pp. 433-471. Discussion, 36, pp. 855-888.

Peyras, L., Royet, P., and Degoutte, G. (1991). "Ecoulement et Dissipation sur les Déversoirs en Gradins de Gabions." *Jl La Houille Blanche*, No. 1, pp. 37-47.

Peyras, L., Royet, P., and Degoutte, G. (1992). "Flow and Energy Dissipation over Stepped Gabion Weirs." *Jl of Hyd. Engrg.*, ASCE, Vol. 118, No. 5, pp. 707-717.

Pravdivets, Y.P. (1992). "Stepped Spillways in World and Domestic Hydraulic Engineering." *Gidrotekhnicheskoe Stroitel'stvo*, No. 10, Oct., pp. 28-32 (in Russian). (Translated in Hydrotechnical Construction, 1993, Vol. 27, No. 10, Plenum Publ., pp. 589-594).

Pravdivets, Y.P., and Bramley, M.E. (1989). "Stepped Protection Blocks for Dam Spillways." *Intl Water Power and Dam Construction*, Vol. 41, No. 7, July, pp. 49-56.

Rozov, A.L. (2003). "Modeling a Washout of Dams." *Jl of Hyd. Res.*, IAHR, Vol. 41, No. 6, pp. 565-577. Discussion: 2005, Vol. 43, No. 4, pp. 435-444.

Schnitter, N.J. (1994). "A History of Dams: the Useful Pyramids." *Balkema Publ.*, Rotterdam, The Netherlands.

Smith, N. (1971). "A History of Dams." *The Chaucer Press*, Peter Davies, London, UK.

TurnbulL, J.D., and McKay, G.R. (1974). "The Design and Construction of Chinchilla Weir - Condamine River Queensland." *Proc. 5th Australasian Conf. on Hydraulics and Fluid Mechanics*, Christchuch, New Zealand, Vol. II, pp. 1-8.

US Department of the Interior (1965). "Design of Small Dams." *Bureau of Reclamation*, Denver CO, USA, 1st edition, 3rd printing.

Visser, P.J., Vrijling, J.K., and Verhagen, H.J. (1990). "A Field Experiment on Breach Growth in Sand-Dykes." *Proc. 22nd Intl Conf. Coastal Eng.*, Delft, Netherlands, B. EDGE Ed., Vol. 2, pp. 2097-2100.

INTERNET REFERENCES

Photographs of stepped spillways	{http://www.uq.edu.au/~e2hchans/photo.html#Step_spillways} {http://www.iahrmedialibrary.net/}
Embankment overflow stepped spillways: earth dam spillways with precast concrete blocks	{http://www.uq.edu.au/~e2hchans/over_st.html}

Air entrainment on chute and stepped spillways	{http://www.uq.edu.au/~e2hchans/self_aer.html}
Gold Creek Dam and its Historical Stepped Spillway	{http://www.uq.edu.au/~e2hchans/gold_crk.html}
Timber crib weirs	{http://www.uq.edu.au/~e2hchans/tim_weir.html}
The Formal water garden	{http://www.uq.edu.au/~e2hchans/wat_gard.html}
Research publications by Hubert CHANSON	{http://espace.library.uq.edu.au/list/author_id/193/} at UQeSpace
UQeSpace open access repository	{http://espace.library.uq.edu.au/}
OAIster open acces catalogue	{http://www.oaister.org/}

Reviewed by Professor Colin J. APELT.

In: Dams: Impacts, Stability and Design
Editors: Walter P. Hayes and Michael C. Barnes

ISBN 978-1-60692-618-5
© 2009 Nova Science Publishers, Inc.

Chapter 5

DISCRETE ELEMENT METHOD FOR THE FORCE DISTRIBUTION IN A CONCRETE-FACED ROCKFILL DAM

Guangqian Wang[1] and Qicheng Sun[2]
State Key Laboratory for Hydroscience and Engineering,
Tsinghua University, China

ABSTRACT

The concrete-face rockfill dam (CFRD) is a simple dam type and has remarkable economical, ecological and environmental benefits, and thus become one of the widely used dam types. As viewed from experiences on construction and operation of a series of 200 m-height scale CFRDs in China, rockfill acts as the main supporting structure and the control of embankment deformation is a critical technology. Physically, rockfill is composed of fragmental materials, such as primary discrete particles. The mass stability is developed by the friction and interaction of one particle on another rather than by any cementing agent that binds the particles together. For dense granular matter, our research team has developed a 3-D numerical code *Tsinghua DEM simulation* (THDEM) by employing rigorous contact mechanics theories. It has the capacity to examine contact force details that are normally inaccessible, and to perform rigorous parametric studies. In this work, by treating a rockfill as an assembly of discrete particles, a small-scale rockfill is constructed with 50,638 poly-dispersed particles, which takes the 233 m high Shuibuya CFRD in China as the prototype. The spatial distribution of interparticle forces and stress propagation within the rockfill are obtained. Concentrated releases of forces are found at specific locations on the slab and the foundation respectively, which indicate the possible deformation of the slab or settlement of foundation. These results would provide an insight into rheological behaviours of rockfill and deformation process of concrete slab face, from a novel viewpoint of heterogeneous distribution of forces instead of averaged stress distribution in rockfill.

[1] Beijing 100084, China, Fax: +86 10 6277 3576. E-mail address: dhhwgq@tsinghua.edu.cn.
[2] qcsun@tsinghua.edu.cn.

Keywords: *Concrete-face rockfill dam; Rockfill; Rheological properties; Discrete element method; Force chain network*

1. CONCRETE-FACED ROCKFILL DAMS IN CHINA

Rockfill has been used in the construction of dams for over 150 years. It was usually dumped in high lifts, with sluicing under high pressure adopted since 1930's. From the late 1960's rockfill has been placed in 1 to 2 m layers and compacted using steel drum vibratory rollers, usually with water added before rolling. The lower compressibility from compaction of rockfill in thin layers resulted in the resurgence in use of the concrete face rockfill dam (CFRD) due to the significant reduction in leakage rate and post-construction deformation. Many developments on the performance of the existing modern CFRD dams, rockfill zoning, dam construction, and seismic analysis have been made in the past decade. One major feature is that the development and use of the smooth drum vibratory roller and design improvements in the cut-off to the foundation, concrete face slabs and joints.

CFRD is a simple dam type and expected to curtailment of construction cost unlike the structure of the conventional rock-fill dam. It is particularly suitable when there is no satisfactory earth available and when a plentiful supply of sound rock is at hand. It could be constructed not only with gravel, low compressive strength rock and on poor foundations, but also with high compressive strength rockfill and on excellent non-erodible foundations. Therefore, very rapid constructions are possible because the process of filling does not have to be interrupted for rolling or other separate compaction operations.

Zoning in CFRDs is a major issue in stability design. Figure 1 shows the zoning of recently completed Shuibuya CFRD in Hu Bei province in China. It is 233m high, the highest CFRD in the world at present. Zones 3A and 3B are the main rockfill zones, and are distinguished by the maximum layer thickness, 0.9m~1.2m for Zone 3A and 1.5m~2.0m for Zone 3B, and maximum particle size allowed up to the layer thickness. Zones 2D and 2E support the concrete slab face, and graded to limit leakage flows in the event of a joint opening or a crack forming in the slab face. Zones 1A and 1B are earth zones to prevent leakage around the plinth, or in event of a leak through the slab face.

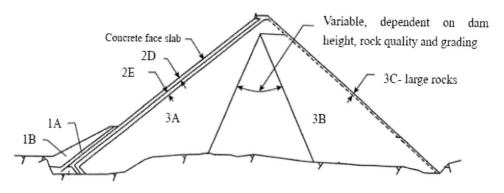

Figure 1. Typical zoning of the Shuibuya CFRD in China.

Table 1 Some CFRDs exceeding 150m in height in China

Name and location	Height(m)/volume (×10⁴ m³)	Construction period	Transition zone: crest/bottom(m)	Dam body dry density (t/m³)	Slab face area (×10³ m²)	Slab thickness crest/bottom (m)
Shuibuya Hu Bei province	233/1526	2002–2007	5/5	2.18	138.4	0.3/1.1
Tianshengqiao (cascade 1) Gui Zhou province	178/1800	1991–2000	5/5	2.10	172.7	0.3/0.9
Sanbanxi Gui Zhou province	185.5/828	2002–2007	6/6	2.17	84	0.3/0.91
Hongjiadu Gui Zhou province	179.5/920	2000–2005	4/10.5	2.19	75.1	0.3/0.91
Tankang Zhe Jiang province	162/948	2005–2009	–	–	95	–
Zipingpu Si Chuan province	158/1117	2001–2006	5/5	2.16	108.8	0.3/0.83
Jilintai (cascade 1) Xin Jiang province	157/836	2001–2006	5/5	2.05	74	0.3/0.85
Malutang Yun Nan province	154/800	Under construction	–	–	–	0.3/0.84

Nowadays, there are many CFRDs higher than 150 m, such as the 185 m high Aguamilpa dam in Columbia completed in 1993 and the 180 m high Tianshengqiao dam in China completed in 1999. Compacted rockfill has also been used in central core earth and rockfill dams since the 1960's and embankment heights have since reached up to 200 to 300 m, such as the 260 m high Chicoasen dam in Mexico completed in 1980, and the 240 m high Guavio dam in Columbia completed in 1989.

In China, more than 150 CFRDs have been built or under construction by the end of 2005, of which 37 are higher than 100 m, 14 higher than 150 m and 15 founded on thick overburdens. Some CFRDs already completed or under construction are listed in Table 1. Especially in the design and construction of the 233m high Shuibuya CFRD, new materials, and new techniques have been applied through scientific research, engineering practice and bold innovation. Important breakthroughs and innovations have been made in relation to the properties of dam materials and test methods, deformation control of dam body, structures and materials for seepage control, dam construction technology and quality control, dam monitoring, and safety evaluation. A full set of key technical procedures and systems for design and construction of high CFRDs have been established.

The economical, ecological and environmental benefits of the Shuibuya CFRD are found remarkable. Compared with an earth core rockfill dam, the total investment on Shuibuya has been reduced by nearly 0.6 billion RMB, the project construction period has been shortened by one year, and the power generation was earlier as well. By the selection of a CFRD for the Shuibuya project, a large quantity of excavated material has been fully used, the excavation of the impervious core material to be used in an earth core rockfill dam was avoided and a considerable area of agricultural land could be saved. In the near future, a few 300m height scale CFRDs are scheduled to be built in the southwestern China where hydro-energy resources are rich, such as the Jinsha River and the Lancang River.

2. Fundamental Study on Rockfill

Rockfill has excellent engineering characteristics, such as good compactness, better seepage ability, high density, low sedimentation, high shear strength and so on. It is widely used in CFRDs, bed of railway and deep foundations. Rockfill is composed of fragmental materials, and the primary particles are discrete from each other. The mass stability is developed by the friction and interaction of one particle on another rather than by any cementing agent that binds the particles together. Two types of rockfill are commonly used in constructions, which are inartificial rockfill such as gravel and scree, and blasted rockfill, e.g. fragmental stone, macadam and moraine soil. Both types are heterogeneous mixture composed of numerous particles whose sizes vary from 10^{-2} cm to 10^{2} cm. The particles have different shapes, dimensions and packing patterns, and thus determine the physical and mechanical characteristics of rockfill. Figure 2 schematically shows that the rockfill could be reasonably resembled with dense compacting coarse particles.

Rheological properties of dam bodies, as one of the fundamental problems, are major concerns in dam stability analysis and design. The related two aspects should be intensively studied respectively, the rheological properties of rockfill, and the deformation of concrete slab face driven by stress from both rockfill and water. Many researchers put forward some stress-strain relations for rockfill through measuring characteristics of particles, acting force between particles and computing distribution of stress based on the elementary theory of statistical mechanics. It is significant and has practical value for CFRDs design and stability analysis. However, satisfactory results have not yet been obtained. The main difficulties are usually from much dispersed mechanical parameters of rockfill, wide size distribution of particles and their heterogeneous location within dams, which causes rheological properties of a rockfill varying spatially and the corresponding uncertain deformations of concrete slab faces.

The State key laboratory for hydroscience and engineering at Tsinghua University in China has a long history carrying out researches on dam design, stability and construction technology. As a new method, a 3-D numerical code *Tsinghua DEM simulation* (THDEM) has been developed for dense granular mechanics studies. The potential value of THDEM is expected to enable to examine data that is normally inaccessible, and to perform rigorous parametric studies. The distinct element method (DEM) was originally developed by Cundall and Strack in 1979 and has been applied to many problems in the fields of powder technology and soil mechanics to reveal the behavior of granular materials. One of the large differences of THDEM from conventional DEMs is that rigorous contact mechanics theories are used in THDEM, such as Hertz, JKR and DMT for appropriate regimes, while simplified spring-dashpot contact modes are often used in DEMs.

Since the rockfill of a CFRD could be treated as an assembly of discrete elastic particles, as illustrated in Figure 2, the position and force on individual particles can be traced by using THDEM, which may have a profound advantage in obtaining the spatial distribution of interparticle force and stress propagation within the rockfill. THDEM is more suitable for the fundamental studies on rockfill than engineering applications. However, THDEM and conventional DEMs have not been received enough attentions and the related basic studies are rather rare.

In Section 3, we describe the THDEM method, related contact mechanics theories and THDEM implementation. DEM models have been developed in recent years and become so complicated as to treat non-spherical (ellipsoidal or polygonal) particles. However, since our study may be an earlier application of DEM to rockfill body packing, we use very simple assumptions: all the particles are spherical and have limited size distribution. Cohesion is not included in the interaction force model, though friction is considered.

Figure 2. Resemblance of a rockfill dam with an assembly of dense coarse particles.

Section 4 introduces a small-scale CFRD prepared in our numerical simulations, which adopts Shuibuya CFRD as the prototype. Size distribution of rockfill in the Shuibuya CFRD is reported, and the parallel graduation method for selecting particle sizes in simulations is employed according to actual size distribution.

Section 5 shows the simulation results, including packing configuration and the internal force chain network, while the later is a main concept for analyzing rheological properties of rockfills. Distribution of inter-particle contact forces is calculated. Concentrated release of forces on slabs and foundation is found and would possibly lead to deformation of slabs or settlement of foundation. The intrinsic mechanism of the force transporting throughout rockfill is discussed in Section 5. Concluding remarks are summarized in the final section.

3. TSINGHUA DEM SIMULATION (THDEM)

The 3-D numerical code of Tsinghua DEM simulation (THDEM) was developed in our research team at Tsinghua University, which follows the basic procedures of discrete element method (DEM) firstly proposed by Cundal and Strack in 1979. However, there is one big difference between THDEM and conventional DEMs, we would say, is that THDEM employs rigorous contact theories while conventional DEMs use simplified spring-dashpot model. The later has to introduce new and unrealistic physical parameters, such as stiffness coefficients for normal and tangential directions. Usually, they require calibrations with experiments.

THDEM simulates the mechanical behaviour of granular assemblies consisting of spherical particles, which is a time-dependent finite difference scheme and used for various simulations. The cyclic calculations of the incremental contact forces and progressive movements of the spheres are performed. For each calculation cycle, the translational and rotational accelerations of each of the constituent particles are given by Newton's second law of motion. Numerical integration is then performed over a small time-step to give updated velocities and displacements for each sphere. The velocities of each particle are used to find the relative approach between contacting particles which is in turn used to calculate the incremental contact forces(normal and tangential) according to contact force-displacement laws. The contact forces are resolved to obtain the out-of-balance forces on each particle, from which new accelerations of each particle are then calculated at the next time-step.

3.1. Newton's Second Law of Motion

The progressive movements of each constituent particle are modeled by Newton's second law of motion, according to which, the motion of a particle over a time step Δt is given by following equations,

$$\text{Translation } F_i(t) + mg_i - \beta_g v_i(t) = m\frac{\Delta v_i}{\Delta t} \qquad (1)$$

$$\text{Rotation } M_i - \beta_g \omega_i(t) = I \frac{\Delta \omega_i}{\Delta t} \tag{2}$$

where $i=1,2,3$ indicates the three components in x-, y-, and z- directions, F_i is the out-of-balance force component of the sphere, v_i is the translational velocity components, m is the mass of the sphere, M_i is the out-of-balance momentum on the sphere due to the contact forces, ω_i is the rotational velocity component, I is the rotational inertia of the sphere, β_g is the global damping coefficient and t is the time. Equation 1 and 2 are solved with finite difference method to give the velocity increments of each constituent particle. The relative displacement increments for each sphere are given as following,

$$\text{Translation } \Delta x_i = v_i \Delta t \tag{3}$$

$$\text{Rotation } \Delta \phi_i = \omega_i \Delta t \tag{4}$$

where Δx_i and $\Delta \phi_i$ are the translational and rotational displacement increments respectively.

3.2. The Interaction Forces between no Adhesion Spheres

Depending on whether the surface adhesion is considered or not, the normal and tangential interactions are modeled according to different theories. In the case of no surface adhesion, the normal force is given by the theory of Hertz and the tangential force is based on the theory of Mindlin and Deresiewicz. In the presence of surface adhesion, the Johnson, Kendall, and Roberts(JKR) theory or Derjaguin, Muller, and Toporov (DMT) theory could be used for the normal force, and the theory of Thornton which combines the work of Savkoor and Briggs, and Mindlin and Deresiewicz is applied for the computation of the tangential force. Tabor showed that both JKR and DMT are limiting cases. Their relations are described in Figure 3. The elastic parameter λ is defined as $\lambda \equiv \sigma_0 \sqrt[3]{9R/2\pi WE^*}$, where σ_0 is the stress at the equilibrium spacing, R and E^* the effective radius of curvature and modulus respectively, W the adhesion energy and P the applied load.

In the case of no adhesion, for two spheres of radii R_1, R_2 and elastic properties E_1, G_1, v_1 and E_2, G_2, v_2, a contact between them exists if their boundaries overlap, i.e. when the relative approach satisfies the following condition,

$$\alpha = R_1 + R_2 - |\vec{r}_1 - \vec{r}_2| > 0 \tag{5}$$

where $\vec{r}_1$ and $\vec{r}_2$ are the centers of the two spheres, as shown in Figure 4.

The normal contact force due to relative approach is, according to Hertz theory

$$N = \frac{4E^*}{3R^*}a^3 \qquad (6)$$

where $a = \sqrt{\alpha R^*}$ is the radius of the contact area which is related to the relative approach of the two spheres α. R^*, E^* are defined as

$$\frac{1}{R^*} \equiv \frac{1}{R_1} + \frac{1}{R_2} \qquad (7)$$

$$\frac{1}{E^*} \equiv \frac{1-v_1^2}{E_1} + \frac{1-v_2^2}{E_2} \qquad (8)$$

Within a time step, if the increment of the relative approach between the two contacting spheres is $\Delta\alpha$, is follows that the corresponding incremental normal force at the contact can be, according to Equation 6, given as,

$$\Delta N = 2aE^*\Delta\alpha \qquad (9)$$

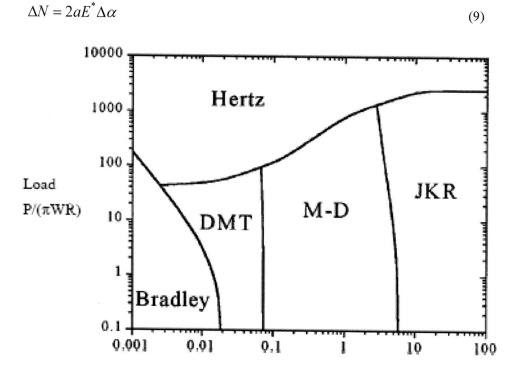

Figure 3. Adhesion map.

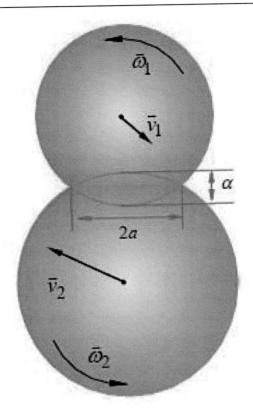

Figure 4. A pair of no adhesion spheres are in contact.

The tangential force at the contact of two spheres is modeled by the theory of Mindlin and Deresiewicz, which predicts that if two contacting surfaces are subjected to an increasing tangential displacement δ, then relative slip is to be initiated at the perimeter and progresses inward over an annual area of the contact surface. The incremental tangential force ΔT due to the incremental tangential displacement $\Delta\delta$ depends not only on the loading history but also on the variation of the normal force. For all cases, the incremental tangential force may be obtained from the following equation,

$$\Delta T = 8aG^{*}\theta_{k}\Delta\delta + (-1)^{k}\mu(1-\theta_{k})\Delta N \tag{10}$$

If $|\Delta T| < \mu\Delta N$

$$\theta_{k} = 1 \tag{11}$$

Otherwise,

$$
\theta_k = \begin{cases} \left(1-\dfrac{T+\mu\Delta N}{\mu N}\right)^{1/3}, & k=0 \\[3mm] \left(1-\dfrac{(-1)^k(T-T_k)+2\mu\Delta N}{2\mu N}\right)^{1/3}, & k=1,2 \end{cases}
\tag{12}
$$

where k=0, 1, 2 denotes the path of loading, unloading and reloading respectively. $\Delta\delta$ is the increment of the relative tangential displacement of the two contact spheres at the contact, μ is the friction coefficient. G^* is defined as

$$
G^* \equiv \frac{2-v_1}{G_1} + \frac{2-v_2}{G_2}
\tag{13}
$$

T_k represents the historical tangential force from which unloading or reloading commenced. It needs to be updated as

$$
T_k = T_k - (-1)^k \mu\Delta N
\tag{14}
$$

to allow for the effect of the variation of the normal force at each time step in Equation 12.

3.3. Time Step

For dynamic processes it is necessary to consider the elastic wave propagation across particles, the time for load transfer from one particle to adjacent contacting particles and the need not to transmit energy across a system faster than in nature. For linear contact stiffness the critical time step is related to the ratio of the contact spring stiffness to particle density. For nonlinear springs (e.g. Hertz) the critical time-step cannot be calculated a priori. However, it was shown that the Rayleigh waves account for 67% of the radiated energy in comparison with the dilatational (7%) and distortional (26%) waves. In the simulations, it is assumed that all of the energy is transferred by the Rayleigh waves. This is a good approximation since the difference between the Rayleigh wave speed and the distortional wave speed is very small and the energy transferred by the dilatational wave is negligible. Upon an application of a force on an elastic body, the Rayleigh waves are propagated along the surface with a velocity of,

$$
v_R = \beta\sqrt{\frac{G}{\rho}}
\tag{15}
$$

where ρ is the density of the material and β is the root of the following equation,

$$\left(2-\beta^2\right)^4 = 16\left(1-\beta^2\right)\left[1-\frac{1-2\nu}{2(1-\nu)}\beta^2\right] \tag{16}$$

From which an approximate solution may be obtained as,

$$\beta = 0.1631\nu + 0.8766 \tag{17}$$

For different material types for the constituent particles in an assembly, the critical time step for the highest Rayleigh wave frequency should be the lowest among those determined by different material types,

$$\Delta t = \pi \left[\frac{R}{0.1631\nu + 0.8766}\sqrt{\frac{\rho}{G}}\right]_{min} \tag{18}$$

It is clear that the detailed treatment of particle interaction in the calculation using rigorous contact theories requires a shorter time step than in a continuum approach. Except that huge computational load in THDEM and other DEMs as well, THDEM exactly enables us to properly calculate particle interactions. Since performance of computers is being drastically improved, and we could effectively handle more than 100,000 of particles in 3D. In this work, we prepare a small-scale rockfill with 50,638 poly-dispersed particles, which takes the 233 m high Shuibuya CFRD in China as the prototype.

3.4. Contact Detection

The simulation of particle systems requires cyclic calculations of particle displacement increments and contact force increments. For the calculation of contact forces between particles, an efficient searching scheme is needed to detect the contacts between particles. For example, the straightforward way to search for the contact particle pairs is to calculate the distances between all the pairs of particles by using Equation 5. Computational time for this procedure is proportional to $\sim O(N^2)$ for N particles, and becomes large when N is large. Because a particle can not make contact with particle which are far away from it, the principle is to search for contacts with only those particle locating in the local vicinity. A more efficient method could be proposed to detect the contact particle pairs. In this study, a spatial cell division method is employed as the contact detection method. The computational space is divided into sub-cells of equal size. The cell size should normally be less than twice the largest sphere diameter so that at least one particle center is packed in one cell. At each time step of the calculation, we make a list of the cells in which particle centers are located. In searching for the particles which are in contact with some a particle, we just search for only the cells containing the particle. This algorithm takes less computational time ($\sim O(N)$) if all the particles have the same size. It should be noted that the procedure used in this study needs much computer memory for the large number of spatial cells.

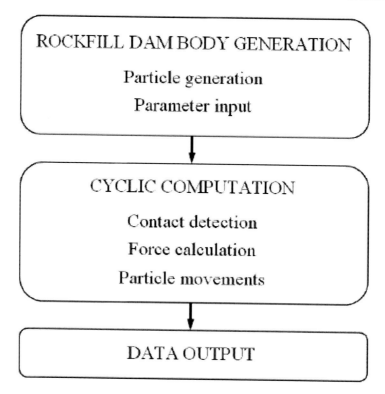

Figure 5. The flowchart for the THDEM simulation.

The implementation procedure of THDEM is as follows. Firstly, the rockfill dam body is filled with randomly packing particles. The motion of each particle at each time step is then calculated. We search for the pairs of particles contacting or overlapping each other in neighbouring sub-cells, and calculate the mechanical interaction forces between them. We also search for the particles in contact with the slabs and foundation, and calculate their interaction forces. Then, we calculate the change in the translational and rotational motions of these particles caused by the mechanical interaction forces by solving the equation of motion for each particle. The mechanical interaction forces between contacting particles are calculated with corresponding contact theories. When the packing process is completed, all information about the particles and walls is output.

4. A SMALL-SCALE CFRD PREPARATION

In this work, we adopt the Shuibuya CFRD in China as a prototype for THDEM simulations. The construction details could be found in Table 1. Due to the limitations of computing capacity, we prepare a small-scale CFRD while keeping the geometry similarity with that of Shuibuya. The particle size in simulations is determined based on actual size distribution by using the parallel graduation method.

4.1. Particle Size Distribution

For a rockfill in which particles are of N different sizes ($N \rightarrow \infty$), the distribution parameters of particle sizes have a probability distribution law. It can be considered that the distribution of particle sizes is continuous. A particle size distributing curve of rockfill could be obtained by dividing the particle sizes of rockfill into several groups, and then measuring the mass fraction and calculating the accumulative mass fraction for each group. Some of the sifted results of rockfill particle sizes in the Shuibuya CFRD are listed in Table 2.

As shown in Figure 1, the main rockfill zones 3A and 3B support the slab through Zones 2D and 2E. The water pressure, weights of the dam body and slab, are transferred through the main rockfill zones onto the foundation. Therefore, the good size distribution in the zones 3A and 3B is required to keep high enough compacting density and modulus to reduce deformation. The particle size distribution at Shuibuya main rockfill zones is shown in Figure 6 (the dashed curve). Since the density distribution of particles was found skewed towards a low-end tail, the grain size distribution would be classified into Talbot distribution,

$$P = \left(\frac{d}{D_{\max}} \right)^{0.45} \tag{18}$$

where P is the percentage of particles smaller than a given size diameter d; D_{max} is the maximal diameter, and power index is proposed to be 0.45 for Shuibuya CFRD.

The solid curve in Figure 6 is the calculated Talbot size distribution. We can see that it is well constituent with the measured size distribution. More details may be found in Table 2.

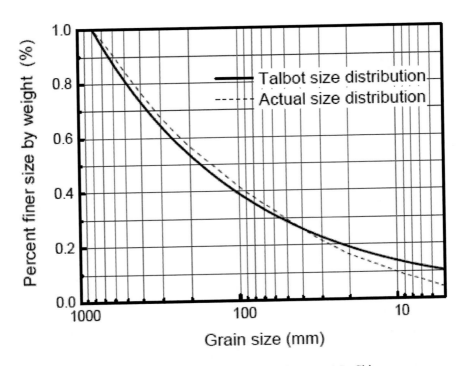

Figure 6. Size distribution of grains in the main rockfill at Shuibuya CFRD, China.

4.2. Parallel Gradation Method

The experimental tests on coarse-grained rockfill or soils always encounter difficulties. It is necessary to remove large particles due to dimensional limitation of laboratory specimens. Another problem encountered in THDEM simulations is that finer particles consume huge computing times, and have to be removed. Therefore, there is a need to assess the effect of particle size and its graduation on the behaviour granular matter in a rockfill. For this purpose, the appropriate method for selecting particle size distribution for simulations is very important.

Three major methods have been widely used, i.e. the matrix model method, the parallel gradation method and the scalping and replacement method. In the parallel gradation method, the oversized particles are scalped and a grain size distribution curve is parallel (in the common semi-log scale as shown in Figure 6 to that of the original sample. The main advantage of this procedure is that the particle size gradation is maintained. However, depending upon the particular characteristics of each rockfill, the mineralogy and hardness of grains, particle shape, and particle roughness, may be different as particle size varies. When these factors are similar for all particles, the parallel gradation method would be an attractive method. One additional limitation arises in those cases where the original coarse material has a considerable content of fines, such that the finer parallel gradation results in sample with an important fines content, for example greater than 10%. The geo-mechanical response of this new batch is strongly controlled by the fines and therefore, it cannot reproduce the behavior of the original coarse material. According to the size distribution of particles in main zones in the Shuibuya CFRD and particles properties, the parallel gradation method is employed in this work..

Table 2. Size distribution in main rockfill zones at Shuibuya CFRD, China

	Mass percentage of size groups										d_{60} (mm)	d_{30} (mm)	d_{10} (mm)
	800~600	600~400	400~200	200~100	100~60	60~40	40~20	20~10	10~5	<5			
Actual distribution	10	12	20	15	11	6	9	8	5	4	210	50	12
Talbot distribution with n=0.45	12	15	20	14	8	5	7	5	4	10	270	56	4.8

Table 3. Size distribution in THDEM simulations in this work

	size groups (mm)				
	600	300	150	80	32
Mass percentage	22	20	15	11	32
Particle number ratio	1	4	11	28	511

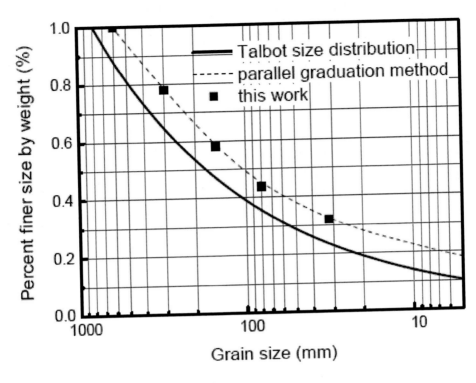

Figure 7. Particle size distribution in this work.

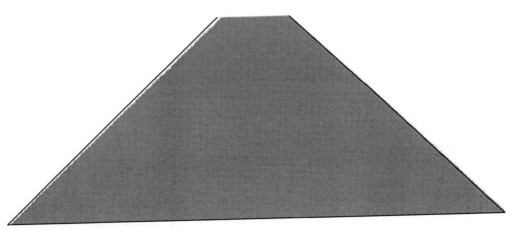

Figure 8. The schematic view of the CFRD in this work. 50,638 poly-dispersed particles are compacted into the space confined by side walls and a foundation wall.

Figure 7 shows the size distribution by using the parallel graduation method (dashed curve). It is parallel to the Talbot curve in the semi-log scale coordinates. To avoid larger number of finer particles, 5 sizes are selected in our simulations. More details may be found in Table 3.

Table 4. Parameter values in this work

Parameters	Values
Rockfill	
height	14.5 m
length at foundation	40 m
upstream slope	1.4
downstream slope	1.4
Particles	
radius	32, 80, 150, 300, 600 mm
density	2.65×10^3 Kg/m^3
Young's modulus	70GPa
Poisson's ratio	0.18
surface friction	0.35
time step	1×10^{-5} s
Slab and foundation	
Young's modulus	13GPa
Poisson's ratio	0.24
surface friction	0.35

4.3. Preparation of a Compact Rockfill

A small-scale CFRD is built in this work, by using the Shuibuya CFRD as the prototype. Both the upstream slab and downstream thick stone layer are replaced with two straight walls. The foundation is represented with a flat wall as well. The geometry details can be found in Table 4. 50,638 poly-dispersed spherical particles are randomly poured into the confined space, and then are compacted with a flat wall on the top moving downwards. The final compact rockfill is shown in Figure 8.

The average coordination number of particles is found 5.2, and the solid fraction is found 0.82. Thus the CFRD in our simulations is dense enough to represent an actual CFRD. Table 4 is construction details for the small-scale CFRD and material properties of particles.

5. FORCE DISTRIBUTIONS IN THE ROCKFILL

The granular matter, including that filling the rockfill in this work, usually displays peculiar mechanical properties. Many of the observed phenomena are still not yet fully understood, and some fundamental mechanical problems still remain open. Examples are dynamic sound propagation, jamming (exceedingly slow dynamics) and creep. Granular matter offers plenty of other puzzles and challenges. Now we should have an insight into the force propagation within the rockfill from a general view.

There are a few words about force and stress in order to illustrate the advantage of THDEM in fundamental study on granular matter. Force is one of basic concepts in classic mechanics, which may lead to a linear acceleration and an angular acceleration. In contrast, stress is a measure of the average amount of force exerted per unit area, which is a basic concept based on the assumption of continuum. The stress concept may be very suitable for describing macroscopically liquid, gas or solid without obvious fracture or interfaces, since it is uniform enough for a small-scale representative volume element. However, for intrinsically discrete solid particles, the effect of heterogeneous distributions of particles becomes greater, and the volume element average method may be not applicable any more, though it is still widely used in present studies on granular matter. We believe that force distribution would be more fundamental than averaged strain-stress relation.

5.1. Force Chain Network

Interparticle forces in granular matter form a heterogeneous distribution of force chains, which might be first termed as *solid paths* by Horne in 1965. Force chains generally are filament-like and parallel to the direction of loads, as shown in Figure 9. The typical length of force chains is around 10 grain diameters. They split and fuse at a variety of angles to form a network. Strong force chains, where the contact force is higher, sparsely distribute throughout the granular matter, which are dominant to mechanical behaviours of granular matter. A larger number of weak chains with lower contact forces exist in the vicinity of strong chains. They usually play an auxiliary role to the stability of strong force chains.

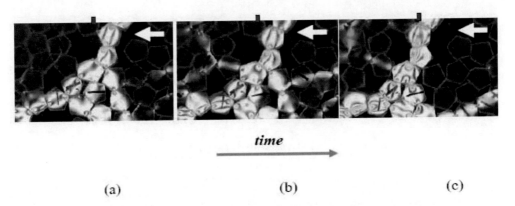

time

(a)   (b)   (c)

Figure 9. Evolution of strong force chains visualized by using the photo-elastic technique. Top layer of particles slides towards left from right as indicated by the white thick arrows.

The chain can only support loads along its own axis: successive contacts must be collinear, with the forces along the line of contacts, to prevent torques on particles within the chain. Neither friction at the contacts nor particle asperity can obviate this, though finite deformability allows small transverse loads to arise. Thus, one would conclude that the occurrence of force chain network is an inherent feature of granular matter, which may substantially contribute to its specific characteristics.

Transmission of large forces is strongly influencing the macroscopic stress. Understanding such forces and their spatial correlations, specifically in response to forces exerting at the system boundaries, represents a fundamental goal of granular matter mechanics. The problem is not only relevant to civil engineering and geophysics, but also to mechanics in explanations of jamming, shear-induced yielding and mechanical responses. This is also what we are much concerned in the fundamental study on rockfill of a CFRD.

In the past few years, a great deal of theoretical and experimental research has been devoted to this domain. However, solving practical engineering problems often requires using phenomenological models, which introduce numerous parameters with no physical meaning. Moreover, a considerable number of constitutive models often contradict with each other, notably in their basic concept. Constitutive models for granular materials based on a micro-mechanical approach remain scarce. In our group, we have proposed a multiscale approach to the mechanics of granular matter, as illustrated in Figure 10. The microscale is related to primary particle and the macroscale is about the granular matter bulk. The contact mechanics between primary particles acts as the basis for the multiscale approach, while the macroscale studies would be related to applications in engineering. Since deriving macroscopic stress equations from known microscopic or grain scale physics has been proven quite difficult, a mesoscale, i.e. the force chain network, should be introduced and would play a vital role in bridging contact mechanics at microscale and macroscopic mechanical properties. The complicated dynamic response to loadings of the force chain network solely contributes to the macroscopic rheological properties of granular matter, which is the main viewpoint of this multiscale approach.

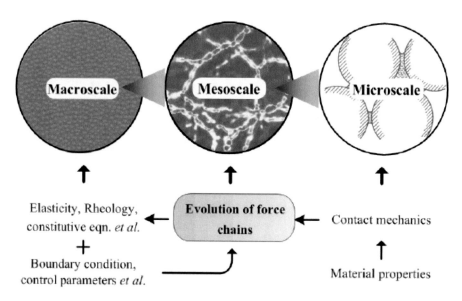

Figure 10. A multiscale method for the study on granular matter mechanics.

Generally, the dynamical system theory and statistical mechanics are two major tools for solving multiscale problems, especially for cases of weak couplings or no couplings. For example, the coupling between different scales is usually analyzed by assuming equal probability: the behaviours on macro-scales are statistical averaging on microscale with a given probability density function. However, scale couplings in granular matter are nonlinear and strong so that they cannot be treated by either statistical methods or small perturbation any further. For example, the microscale particle properties determine formation and stability of a force chain, while the interactions between force chains favor the macroscopic stress, as illustrated in Figure 10. The unifying of the two correlations may eventually constitute a generalized theory for granular matter mechanics. Due to constrain in experimental measures, computer may be merely the last useful tool to investigate granular matter mechanics.

5.2. Force Chains in the Rockfill

In this work, adhesion between contacting particles could be neglected according to size distribution and loadings. According to the adhesion map as shown in Figure 3, the normal force is given by the theory of Hertz and the tangential force is based on the theory of Mindlin and Deresiewicz.

Figure 11 shows the contact force chain network throughout the CFRD. Contact force F is classified into three ranges, i.e. $F<800N$, $F \in [800N, 4,800N]$ and $F>4,800$ N. The chain thickness is proportional to the magnitude of local contact forces. We also draw force chains in different colour for easily viewing their difference. For higher F ($>4,800N$), the chains, displayed in black colour, are sparsely distributed within the rockfill, especially below the middle part. For small F ($<800N$), the chains, displayed in gray colour, exist nearly everywhere in the rockfill. The occurrence of force chain network may be an inherent feature of granular matter, which may substantially contribute to its specific characteristics.

Figure 11. Force chain network within the CFRD after compacting. Black chains correspond to greater contact force ($>4,800$ N); dark gray chain correspond to contact forces greater than 800 N and less than 4,800 N; gray chains correspond to contact force less than 800 N. The thickness of chains is proportional to the magnitude of the forces.

Figure 12. Force chain network within the CFRD before compacting. Black chains correspond to greater contact force (>4,800 N); dark gray chain correspond to contact forces greater than 800 N and less than 4,800 N; gray chains correspond to contact force less than 800 N. The thickness of chains is proportional to the magnitude of the forces.

Since the rockfill is compacted by moving downwards a wall on the top to represent the rolling process in constructions of CFRDs, we notice that resulted strong force chains below the top surface are nearly parallel to vertical compacting direction. It further indicates that force chains could only support loads along its own axis, though finite deformability allows small transverse loads to arise. How to describe the network style still remain open so far, although fractural dimension analysis and self-similar network method have been tried.

To compare the variation of force chain networks after compaction, Figure 12 is network of the loose rockfill before compacting. It is clear that weak force chains (in gray colour) are mainly locating in the upper part; while in the compacted CFRD, the weak force chains distribute nearly uniformly throughout the whole CFRD. Strong forces are not quite different between these two cases, except more strong forces in upper part of the compact rockfill. In the following, we would analyze the magnitude distribution of inter-particle force, force releases on slabs and foundation respectively.

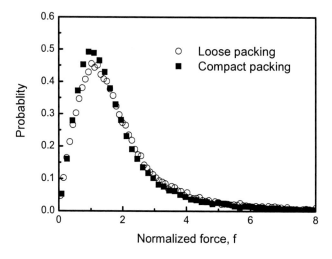

Figure 13. Distribution of inter-particle forces in the CFRD.

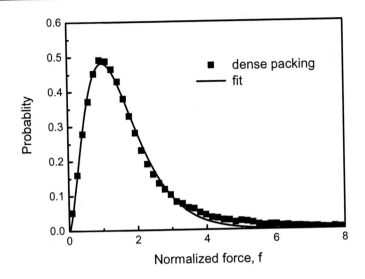

Figure 14. Inter-particle force distribution.

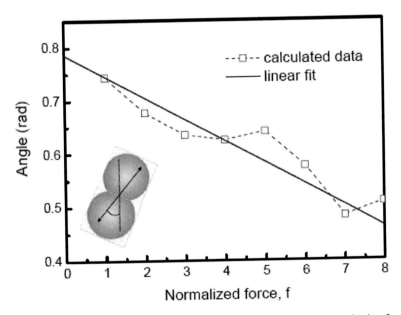

Figure 15. Variation of the angle between contact force and gravity with the magnitude of contact forces. The inlet shows the angle.

5.3. Inter-particle Forces

Figure 13 is the probability distribution of normalized inter-particle force f, where f=F/<F> and <F> is the mean force. From Figure 13, we notice that the force distribution is basically the same, regardless of loose rockfill or compact rockfill. Both reach a peak value of probability around the mean value (i.e. f=1), and have similar long tails.

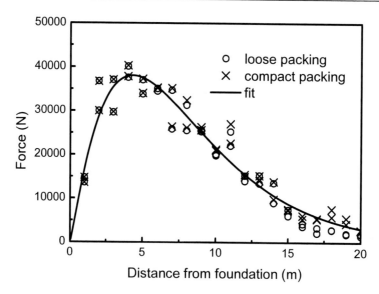

Figure 16. Forces released on the slab. Crosses denote forces before compacting, while circles denote forces after compacting.

This interesting phenomenon has been reported for many other granular static packings, and may be one of the major characteristics of granular matter. For example, relying on DEM simulations, the statistical distribution of contact forces inside a confined packing of circular rigid disks with solid friction was studied by Radjai et al. in 1996. They found the following: (1) The number of normal and tangential forces lower than their respective mean value decays as a power law, i.e. $P(f) \propto f^{\alpha}$ for $f < 1$. (2) The number of normal and tangential forces higher than their respective mean value decays exponentially, i.e. $P(f) \propto e^{-\beta f}$ for $f > 1$, and $1.0 < \beta < 1.9$. (3) The ratio of friction to normal force is uniformly distributed and is uncorrelated with normal force. (4) When normalized with respect to their mean values, these distributions are independent of sample scale and particle size distribution.

In this work, we propose to use the following function for the case of compacted rockfill,

$$P(f) = k f^{\alpha} e^{-\beta f} \tag{19}$$

where $k=2.887$, $\alpha=1.789$ and $\beta=1.794$ which locates in the range of (1.0, 1.9). As shown in Figure 14, the fit curve is well consistent with numerically measured data as $f < 3$, and only slightly diverges as $f > 3$.

Basically, particles are randomly dispersed in the rockfill. The directions of inter-particle forces are divergent from gravity. The angle between contact force direction and gravity is defined as below,

$$\theta = \sum_{\{ij\}} \theta_{ij} \tag{20}$$

where $\theta_{ij} = \arccos\left[\vec{r}_{ij} \cdot \vec{g}\left(\left|\vec{r}_{ij}\right|\left|\vec{g}\right|\right)^{-1}\right]$ for $\vec{r}_{ij} \cdot \vec{g} > 0$, else $\theta_{ij} = \pi - \arccos\left[\vec{r}_{ij} \cdot \vec{g}\left(\left|\vec{r}_{ij}\right|\left|\vec{g}\right|\right)^{-1}\right]$. i and j denote a pair of contacting particles.

Figure 15 shows the schematics of the angle θ, and the variation of the angle between contact force and gravity with the magnitude of contact forces. From Figure 15, we notice that the angle is close to $\pi/4$ at smaller contact forces. As contact forces increasing, the angle decreases. If the loading is much greater, the angle may reduce to zero, which indicates the axial direction of force chain would be parallel to the greater loading.

5.4. Forces on the Slabs

The force exerting on slabs is one of major concerns in the stability analysis and design of CFRDs, because possible concentrated release of force onto slabs would certainly deform local slab faces. Figure 16 is the force distribution along the slab from the foundation to the crest. We can see that force increases sharply from the foundation until reaching a peak value of 40,000N at 5m, and then slowly decreases until the crest. It indicates that the local slab at 5m starting from the foundation is easily to be broken, i.e. about at the place of 1/4 slab length. Under real circumstances, water pressure should be considered to eventually determine the possible maximal force position, and the easily breaking place of slabs is slightly different from the place of 1/4 slab length, such as at the place at 1/3 slab length as reported in literatures. One interesting thing is that this tendency is nearly the same for both loose packing and compact packing of rockfill. Its explanation may involve the force transporting mechanism of granular matter, which would be discussed in the next section.

We propose a function to fit the data,

$$F = 21929.7s^{1.189}e^{-0.276s} \tag{21}$$

Where s is the distance along the slab starting from joint between slab and foundation.

5.5. Forces on the Foundation

Another major concern of CFRD stability is the settlement of foundation. One usually carefully measures mechanical properties of foundation and then takes necessary construction techniques. In this work, we would say that the anisotropic release of rockfill weight, water pressure and possible external loadings could lead to local settlement of foundation. Figure 17 is the contact force propagated onto the foundation. We can see that a maximum exists around in the center of the foundation so that the center part would settle firstly. As the width less than 5m or greater than 35m, i.e. the two ends of the foundation, forces changes greatly, which could cause great shear deformation gradation of the rockfill. Meanwhile, we notice that there is a large increase around the central part after the rockfill is compacted.

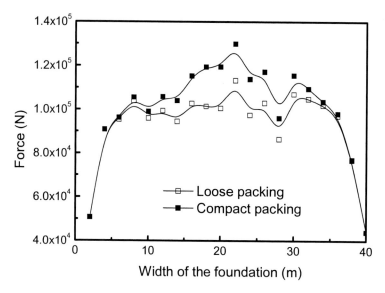

Figure 17. Forces released on the foundation.

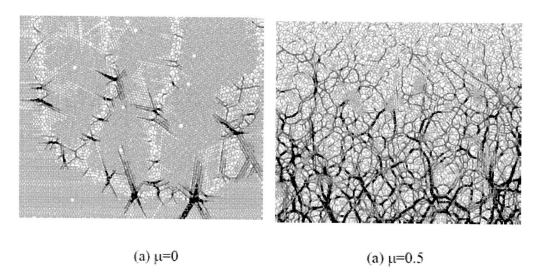

(a) μ=0   (a) μ=0.5

Figure 18. The variation of force chain networks with particle friction μ. (a) μ=0, and (b) μ=0.5. (b) is a part from the Figure 11 and (a) is a new simulation with μ=0.

Understanding of Figure 16 and 17 may involve the mechanism of the transmission of force through granular matter, which is an essential question for researchers investigating the stability of buildings, silos and slopes — particularly for predicting failure and avalanches.

What is the form of an equation that describes the propagation of information and, more specifically, the force, or stress, distribution in granular matter? For granular materials such as sand, soil or snow, material disorder, anisotropy (direction dependence of properties) and friction have to be considered. In a fluid, the stress increases linearly with depth. But in a silo, the stress in the granular material saturates at a certain level that is independent of the depth, because static friction (not present in a fluid) causes the weight of the sand to be partially transferred to, and carried by, the silo's walls.

When there are no side walls, as in the case of a conical sandpile, the situation is even less clear. One would expect the stress to be maximal under the apex (the highest point). Surprisingly, some experiments reveal that the stress actually shows a local minimum — a 'dip' — right under the apex. This observation is one of the reasons behind the revival of interest in the problem of force (or stress) propagation in granular materials. Although the dip cannot be explained using the simplest elasticity theory, wave-like models readily predict it. But the dip can also be reproduced by including anisotropic elasticity, for instance. Invoking special boundary conditions, such as a deflection of the supporting ground or a rough floor, allows elastic models also to mimic the dip.

Furthermore, it turns out that the existence of the dip depends on the structure of the sandpile and the way it was constructed. Thus, the problem also concerns the interplay between extrinsic boundary effects and intrinsic granular-material behaviour.

5.6. Effect of Friction on Force Chain Structure

Figure 18 shows that the simple act of removing surface friction can cause a great variation of force chain pattern. From Figure 18(a), we can see that the regular, lattice-like structure of particle packing occurs due to $\mu=0$. The particle assembly is thus divided into a few parts and among the fractures some particles are squeezed to form strong force chains. While in separated parts, the connections of particles are weak. Therefore, friction is vital to the strength of force chains. Thus, while it is convenient to remove surface friction in many analyses or computer simulations, it is removing essential physics from the problem and will lead to erroneous behavior. At the same time, the importance of surface friction may be dwarfed by the importance of particle shape. Angular particles have a greater tendency to lock together and will thus both more easily form force chains and form stronger force chains. The effect of particle shape has yet to be explored in detail. However, results on frictional spheres have some value. One can at least make frictional spherical particles; frictionless particles, spherical or otherwise do not appear in nature.

6. CONCLUSIONS

The concrete-face rockfill dam is feasible dam type for hydro-power station constructions in southwestern China, especially under the circumstance of poor foundation and a plentiful supply of sound rock. As the schedule, a few 300m height scale CFRDs are scheduled to be built there along Jinsha River, the Lancang River and so on. Rockfill is the supporting structure and the control of embankment deformation is the core of key technology for construction of extra high CFRDs. The fundamental studies on the mechanical properties of the rock-fills should be strengthened for construction analysis of extra high CFRDs. The THDEM code we developed would be an appropriate tool. It could provide force distributions among particle and releases on both slabs and foundation. The obtained results in this work have shown this point.

In the next work, we will realize parallelization of our THDEM code so that larger-scale CFRDs could be numerically studied and the actual size distribution would be used directly. By then, the simulation results could be compared with experimental measurements. From the force release on slabs and foundations, helpful suggestions can be proposed to avoid possible deformation of slab, and possible settlement of foundation. Meanwhile, we will investigate force chain network and its contribution on the macroscopic mechanical properties of rockfill.

ACKNOWLEDGEMENTS

The authors acknowledge the support of the National Key Basic Research Program of China (Grant No. 2007CB714100).

REFERENCES

[1]   Campbell C S. Stress controlled elastic granular shear flows. *J. Fluid Mech,* 2005, 539: 273–297

[2]   Clements R P. Post-construction deformation of rockfill dams. *J. Geotech. Eng,* 1984, 110(7): 821–840

[3]   Cooke J B. Progress in rockfill dams (18th Terzaghi lecture). *J. Geotech. Eng,* 1984, 110(10):1383–1414

[4]   Cooke J B. Rockfill and the rockfill dam. Proc., Int. Symposium on High Earth-Rockfill Dams, G. Jiang, B. Zhang, and M. Qin, cds., Beijing, Chinese Society for Hydro-electric Engineering, 1993, Beijing, 1–24

[5]   Cooke J B. The concrete face rockfill dam. Non-Soil Water Barriers for Embankment Dams, 17th Annual USCOLD Lecture Series, San Diego, United States Committee on Large Dams, 1997, Denver, 117–132.

[6]   Cooke J B. The high CFRD dam. Proc., Int. Symp. on Concrete Face Rockfill Dams, J. Barry Cooke Volume, R. T. Mori, J. A. Sobrinho, H. H. Dijkstra, J. Guocheng, and L. Borgatti, eds. International Commission on Large Dams, 2000, Paris, 1–4

[7]   Cundall P A, Strack O D L. A discrete numerical model for granular assembles. *Geotechnique,* 1979, 29(1):47–65

[8]   Duran J. Sands, Powders and grains: an introduction to the physics of granular materials. Springer-Verlag, Heidelberg, Germany. 2000

[9]   Goldenberg C, Goldhirsch I. Force Chains, microelasticity, and macroelasticity. *Phys. Rev. Lett,* 2002, 89: 084302

[10]   Horne M R. The behaviour of an assembly of rotund, rigid cohesionless particles—I, II. *Proc. R Soc.* London A, 1965, 286: 62–97

[11]   Johnson K L. Contact Mechanics. Cambridge University Press, Cambridge, England, 1985

[12]   Luding S. Granular media: Information propagation. Nature, 2005, 435: 159–160

[13]   Ma H Q, Cao K M. Key technical problems of extra-high concrete faced rock-fill dam. Sci. China Ser. E: *Tech. Sci,* 2007, 50:20-33(Supp.I)

[14] Mindlin R D, Deresiewicz H. Elastic spheres in contact under varying oblique forces. *J. Appl. Mech*, 1953, 20: 327–344

[15] Radjai F, Jean M, Moreau J J, Roux S. Force distributions in dense two-dimensional granular systems. *Phys. Rev. Lett.* 1996, 77(2):274–277

[16] Saboya F J, Byrne P M. Parameters for stress and deformation analysis of rockfill dams. Comput Graph Image Process, 1993, 30, 690–701

[17] Sherard J L, Cooke J B. Concrete-face rockfill dam: I. Assessment. *J. Geotech. Eng,* 1987, 113(10): 1096–1112

[18] Savkoor A R, Briggs G A D. The effect of tangential force on the contact of elastic solids in adhension. *Proc. R Soc. Lond.* A, 1977, 356:103–114

[19] Thornton C. Interparticle sliding in the presence of adhesion. J. Phys. D: *Appl. Phys,* 1991, 24:1942-1946

[20] Tan J Y. Development of dam engineering technology in our country. *Water Power,* 2004, 12: 60–63

[21] Unertl W N. Implications of contact mechanics models for mechanical properties measurements using scanning force microscopy. *J. Vac. Sci. Technol.* A, 1999, 17(4): 1779–1786

[22] Yang Z Y, Zhou J P. Prospects for Construction and technology of extra high concrete faced rock-fill dams. *Water Power,* 2007, 1: 64–68

In: Dams: Impacts, Stability and Design
Editors: Walter P. Hayes and Michael C. Barnes

ISBN 978-1-60692-618-5
© 2009 Nova Science Publishers, Inc.

Chapter 6

EROSIVE AND ENVIRONMENTAL IMPACT OF HYDROLOGICAL CORRECTION CHECK DAM

F. Belmonte Serrato[1] *and A. Romero Díaz*[2]

Department of Geography, University of Murcia, Spain

ABSTRACT

The hydrological correction check dams, transversely built in the riverbeds, have as a main aim the retention of sediments arriving to the reservoir, delaying their silting and expanding their lifetime.

In this research we deal with two aspects related to the environmental impacts that trigger the construction of hydrological correction check dams: the erosion caused within the riverbeds and the ploughing and soil movements made by the opening of paths in order to grant access to the check dams.

Downstream of these check dams, an incision is produced as a result of the slope breaking provoked by the check dam. The length of the stretches cofferdamed by erosion downstream of these check dams in the riverbeds studied, fluctuates between 50 and 150 meters from the bottom of the check dam. It has been calculated that the sediments evacuated in these stretches represent between 10 and 15 per cent of the accumulated deposits in the check dams located further down, reaching in some cases 50 per cent. This erosion reduces considerably the lifetime of the check dams and question, in some cases, their own usefulness.

In the other hand, the access paths to the check dams´ construction points have an environmental cost, which in some cases, go against the aim of such constructions and are difficult to justify given the little sediment accumulation capacity of the check dam, or the low rate of erosion of the gully where it is being built. The study carried out in the basin of River Segura, located in the southeasterm of Spain, shows that the relation between the soils removed in the access paths and the soil retained is approximately 10 per cent, although the fluctuations are between 7 and 36 per cent. Besides, from these paths and at the heights of each check dam, some others are constructed that are directly going to the basin of the check dam, with slopes that in many cases, go over 100 per cent

---

[1] E-mail: franbel@um.es.

[2] E-mail: arodi@um.es.

and finally become a source of "non natural" sediments added to the already mentioned erosion of the riverbeds. Finally, the ploughed area with bushes and pine elimination has an approximately average of 0.7 hectares per check dam, reaching 1.6 in one of the gullies.

Notice that, the erosion added effects in the riverbeds and the erosion in paths that go directly to the basin of the check dam can represent as an average, 20 per cent of the sediments accumulated in each check dam. This erosion will not be produced if check dams had not been built, and apart from this, it is necessary to get rid of almost a hectare of bushes per check dam, for building access paths, eliminating in many cases, protected species and been certain that the recovery of these bushes in a semiarid environment as the one in the river Segura basin will be produced in the long term or even it will not happen at all, the check dams construction in certain gullies is not justified.

In short, before the construction of check dams is essential to consider several aspects, because in some occasions the environmental and erosive impact produced by the construction itself can be higher to the benefits we pretend to obtain.

Keywords: *Hydrological correction works, check dams, sediments accumulated, erosion, river channels, useful life, southeastern Spain*

1. INTRODUCTION AND AIMS

The construction of hydrological correction check dams usually is done over basins all over the world and in Spain are quite common (Varela Nieto, 1999; Martinez de Azagra *et al.*, 2002), but where we can find the highest number of them is in the semi arid areas, due to the scarce vegetation cover, torrential rainfall and, as a result, the intensity of the erosion and sedimentation processes.

Figure 1. Gabion check dam in a gully over the River Quipar basin, tributary of River Segura in the Southeastern of Spain.

Figure 2. Silted check dam in a gully over the River Quipar basin. The difference in height between the spit of sediments and the riverbed downstream is 4.90 meters.

Spain with a total of 1.401 reservoirs (MMA, 2006) is the European country with the biggest number of dams and the fourth in the world ranking (WCD, 2000). However, the important investment done in their construction has an implacable enemy, the eroded sediments in their watershed basins that are deposited in the vessel of the check dams reducing this way their capacity. The current average loss of capacity in the Spanish reservoirs is 10 per cent or 0.5 per cent per year. (Avendaño et al., 1997). For avoiding the reservoirs silting is necessary the construction of hydrological correction check dams for retaining sediments and expanding their lifetime (Romero Diaz et al., 2004).

In the River Segura basin, located largely in the semiarid environment, currently has 47 reservoirs, 14 of them are of recent construction, as a defence from flash floods, but built at the beginning of the XX century, they have more than 40 per cent of their basin silted (Romero Diaz et al., 1992; Sanz Montero et al., 1998). That is why the number of check dams built in all rivers and watercourses over the main riverbed is very high. Nowadays the estimated total number of check dams over the River Segura basin is approximately 4000 (figure 1).

In one of the most important tributary rivers within the river Segura basin, river Quipar, 425 check dams have been built. The main aim of these infrastructure works has always been the avoiding of the Alfonso XIII reservoir silting, located in the mouth of the river, before its convergence with river Segura. Such reservoir with a capacity of 42 cubic hectometres in 1916 ended up with 14.2 cubic hectometres in 1976 (Romero Diaz et al., 1992). Because of this fact the Confederación Hidrográfica del Segura (CHS) carried out two hydrological correction plans in this basin, the first one in 1962 (CHS 1962); and the second one in 1996 (CHS 1996); nowadays, the CHS is carrying out a third project.

Table 1. Check dam Operational Stage for the total extension of the basin

	Check dams built 1962	Check dams repaired 1996	Check dams enlarge 1996	Check dams built 1996	Total check dams 1962+1996
Number of check dams	191	58	100	76	425
Silted	73	3	53	14	143
Operational	73	55	47	62	237
No data	45	0	0	0	45

Under the magnitude of the basin semiarid erosion processes, as the one in river Segura in the Southeast of Spain, and the speed with which the reservoirs are losing their capacity, it is urgent to stop or slow down these effects through measures such as the forest hydrological restoration. But in too many cases, the rushes, lead to the execution of these works without carrying out the prospective works of the hydrological geomorphologic or erosive consequences, that those works would have in their basins and in the immediate surroundings, and, even less environmental impact studies generated by the necessary works for the construction of such check dams, by the way, mandatory in Spain only from 1986.

In the hereby studied basin, the one in river Quipar (tributary of river Segura), the first construction works were done by the CHS in 1962. In this project, the hydrological and forestation correction works are being done in 40 waterways and gullies, building 349 check dams and a big number of stone walls. In that time, the environmental impact study carried out was not compulsory, so, the works were executed without previous environmental impacts studies.

In 1996 (CHS 1992), a new project is done for the same basin, repairing 58 check dams, rising the height of 101previously built (those already silted) and building 76 new ones (table 1). In this case, the correspondent project is accompanied by the Environmental Impact Study, although inside, it is remarked the no need of it, in what is related to the check dams construction, under the Royal Decree Law 1301/86 June the 28th.

All the check dams built in the River Quipar basin have been the matter of study within a research project, whose most significant results have been published in Romero Diaz *et al.*, (2007).

In the researches that have been done along the last few years over basins affected by hydrological correction works (Garcia Ruiz and Puigdefabregas, 1985; Martinez Castroviejo *et al.*, 1990; Brandt 2000; Conesa Garcia *et al.*, 2004), the first consequence of the transversal check dams construction in the riverbeds is the breaking of the riverbed balance slope as a result of the progressive change of the local base level, as long as the height of the sediments wedge rises (figure 2).

The search for a new balance profile is prolonged along the lifetime of the check dam, because while the check dams is accumulating sediments, produces at the same time a continuous change of the local base level. This represents that the reduction in the slope of the riverbed in the area of the sediments spit, provokes an erosion or incision in the check dam's riverbed downstream that prolongs hundreds of metres even up to the point of reaching the sediments spit of the check dam located further down (figure 3).

Figure 3. Hydrological correction check dam built in 1962 and enlarged in 1996, with indication of the sedimentation and erosion areas. It is perceivable that the erosion in the riverbed down the river is promoting the breaking of the reservoir itself.

Figure 4. Access path to the basin in a gully for easing the construction of a check dam with 45 ° degrees steepness.

Figure 5. Emptying section over the original level in a stretch of access to the basin for the construction of a check dam.

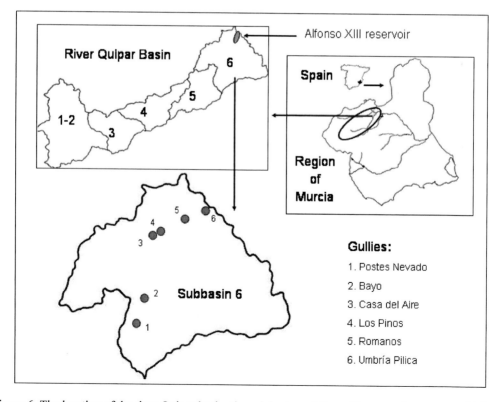

Figure 6. The location of the river Quipar basin, the sub basin and the gullies over the ones on which the study has been based.

As in the same riverbed several check dams are built, evidently, the evacuated sediments as a result of the riverbed erosion, ended up contributing to the silting of the check dam downstream and, as a consequence, reducing its lifetime. Evaluate the volume of eroded sediments in the riverbeds as a consequence of this process, and to determine its importance in the lifetime reduction of the check dams are the first of the aims of this research.

On the other hand, the construction of these works in gullies, in most of the cases, with very difficult access, requires a previous path surfacing for easing the transport of materials and machinery needed in the chosen construction point, with the obvious geomorphologic, erosive, hydrological and ecological impact. In some of the oldest check dams, access path constructions are not observed, but in the new ones built in 1996, in the oldest repaired and the expanded ones, the access paths built for giving way to the heavy machinery to the very limit of the check dam (in the riverbed itself), have a considerable magnitude.

The general overview done over all the check dams in this basin, has allowed us to observe that neither the ploughed areas for the construction of these access paths, nor the stretch of access to the bottom of the riverbed, that in some occasions have steepness higher than 100 per cent (figure 4) and in some other cases cofferdams over the original surface of more than 2 meters high (figure 5) have been reforested again. That is why the evaluation of the ploughed area and the removed soil measure for the construction of these paths, represents the second aim of this research.

2. AREA OF STUDY

The river Quipar basin with an extension of 826 square kilometres is a tributary of the river Segura through its right bank and it is placed in the occidental centre of the Region of Murcia, in the southeasterm of Spain (figure 6).

Its altitude goes from 200 to 1900 metres, but 50 per cent of its area is located well above 800 metres. The limestone and dolomite domain the upper part of the basin, while in the middle and low there is a wide lithologycal variety, in which we have limestone, marls, limestone–marls, marls-gypsums, sands, sandstones and quaternary formations. The predominant soils, developed over these lithologies are different kinds of *Regosols, Leptosols, Calcisols* and *Fluvisols*. (Alonso *et al.*, 2003)

The vegetation is mainly made of xerofiles conifer woods such as the *Pinus halepensis* and some other brushes such as: juniper, rosemary, ~~brushwood, sweet basil,~~ esparto grass, and thyme areas. The protection of the soil by the vegetation tends to be less in those that are more in need, such as the soils with marls like lithology in which the vegetation is usually made of brushes with little demeanour or low density.

The annual average rainfall is 287 millimetres in the lowest part of the basin and around 400 millimetres at the source, with a high inter-annual irregularity. The average temperatures vary from 12 to 16 degrees. According to the Turc classification, the whole basins possess a semiarid climate (Martinez Lloris *et al.*, 2001).

Table 2. Check dams built in the River Quipar basin as in sub basins

	Built	Repaired	Enlarge	New	Total
	1962	1996	1996	1996	1962 and 1996
Sub basin	N°	N°	N°	N°	N°
1,2,3 (Source)	0	0	0	13	13
4 (High – middle stretch)	38	13	20	16	87
5 (Mid – low stretch)	60	21	16	15	112
6 (Mouth)	93	27	61	32	213

Table 3. Erosion rates as in sub basins

Sub basin	Erosión rate (T/hec/year)	Number of check dams used
1.2.3 (Source)	0,71	13
4 (Mid – High stretch)	3,39	44
5 (Mid – Low stretch)	2,64	43
6 (Mouth)	5,29	95
Average	3,95	195

3. METHODOLOGY

The river Quipar basin is divided in six sub basins (according to the CHS denomination) of different lithological and topographic characteristics. The basins five and six are the basins with the highest number of check dams (table 2), and from these two, the most difficult one is the sub basin 6, it is the closest to the reservoir, where the predominant lithology are marls, limestone–marls and clays with a very little development of the vegetation cover, with a rainfall fluctuating between 250 and 350 millimetres and in which until the last half of the last century the intensive harvesting of esparto grass was very common in the area, fitting out the land for such activity. The end of the esparto grass trading brought the abandoning of these areas and the speeding up of the erosion processes. The basin number six in where the biggest number of check dams have been built and where the erosion rates are higher (table 3), over 5 tonnes per hectare per year as an average (Romero Diaz *et al.,* 2008) That is why the chosen basin for this research has been number 6.

3.1. Erosion within the Basins

In the six gullies selected of basin number six, we measured, downstream of each check dam, the length of the stretch modified by erosion in the riverbed as a result of the breaking of the slope that represents the construction of an obliquely check dam in the riverbed. The identification of these stretches has been made possible thanks to the presence of lateral deposits (little terraces) corresponding to the base level prior to the check dam construction (figure 7). In some cases, within narrow stretches, the erosion has affected the whole oblique section, in a way that it has been impossible to identify the previous base level deposits, so all these check dams have been ignored.

After measuring the modified stretch three oblique profiles were made, at the beginning, in the middle and at the end of each stretch. The profiles were levelled taking as a base the one from the check dam that approximately corresponds to the base level in the point of the riverbed before the check dam construction.

The profiles are made placing two metal bars, one in each side of the check dam, in which a mark is done corresponding with the base level of the check dam, to which we awarded value 0. After, between the two bars we tighten a rope in which we have previously placed longitudinally, every 50 metres, marks, and as a last procedure, we measured the vertical length between the longitudinal marks and the floor (figure 8). This way, a transversal section of the riverbed is obtained with a 50 centimetres resolution (the resolution can vary in respect to the accuracy we want to obtain from the transversal section).

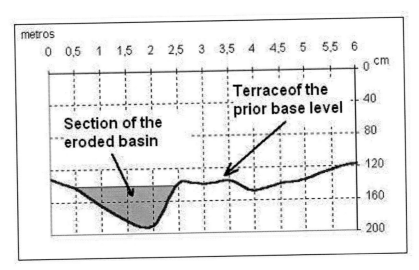

Figure 7. Riverbed section eroded down the river and base level prior to the building of a check dam in the Los Romanos gully. The transect is found 35 metres downstream of the check dam, and the 0 corresponds to the base level of the same one.

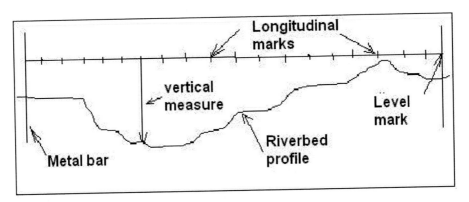

Figure 8. Procedure diagram for obtaining the profile of a riverbed.

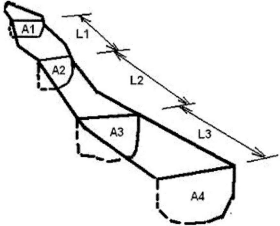

Volume = $\sum$ (A1 + A2 *L1 / 2) +(A2 + A3 *L2 / 2) +.

Figure 9. Transversal section method formula.

Figure 10. Initial emptying section and gully previously developed of approximately 1.5 and 2.5 metres of depth, in an access path to the riverbed for a check dam construction.

The emptying average section obtained from the three profiles by the length of the modified stretch; provide us with a very approximate volume of the sediments exported by the erosion in the riverbed downstream of each check dam.

The volume of exported sediments by erosion in the affected sections of the riverbed was carried out through the procedure of transversal sections (figure 9).

After, the surface of each the check dam's sediments wedge has been measured as well as we have estimated the volumes according to the most suitable geometrical figure, in this case, through the trapezoidal base pyramid volume formula, delimited by the extension of the sediments over the check dam, their depth right in the check dam base, taking as base level, the height of the check dam up to the sediments and the length of the spit of the sediments from the check dam to upstream the river (Hernandez Laguna *et al.*, 2004)

$$V= 1/3 * B * H$$

where: V=Sediments volume;

      B=Area of the base matching with the check dam

      H=Length of the sediments spit

      The perimeter of the sediments spit was obtained through geo-referencing points taken every two metres thanks to the GPS, from where we got its maximum length.

3.2. Environmental Effects of the Paths Opening

For the paths, over a map 1: 5000 the length of the paths, opened precisely for the construction of the check dams, have been measured. Afterwards, on the field, the stretches that didn't appear in the map have been completed, and those that have been proven as already existing previously for giving access to agricultural properties have been ignored.

In all the stretches, identified as open paths for the check dam's construction, we have obtained the average width and the removed soil depth. With this, we have calculated the volume of earth removed and the total of surface ploughed, relating both parameters with the number of check dams and their sediments storing capacity.

Finally, the soil removed, because of the erosion in irrigation ditches and gullies in direct access paths to the riverbed that ultimately have finished sedimented in the vase of the check dam, have been measured (figure 10).

4. RESULTS

4.1. Erosion within the Riverbeds

a. Sediments in the Check Dams

In the gullies studied, 57 unstilted check dams have been selected and all of them with the same years in service. The accumulated sediments as a whole represent 69.140 cubic metres (table 4), for a total area of 3.042 hectares. In the analysis done, no relation between the size of the basin and the retained sediments has been appreciated, this is due to the wide diversity between the big quantities of retained sediments in the 19 check dams of "Los Romanos gully" and those retained in the "Postes del Nevado gully" that has a three times bigger basin area. Neither the relation between the number of hectares per check dam and the accumulated sediments has proven any significant difference. However, we can observe a clear relation between the accumulated sediments volume and the existing number of check dams in each gully (figure 11). This relation, logical by the way, says nothing about the sediments production in gullies, it only shows that the quantity of sediments contained in the check dams is very similar in all of them, because all of them have the same operational life, we can finally say that the accumulation rate is also very similar too.

This lead us to believe that as the accumulation of sediments in the check dams does not have any relation with the size of the basin, at least part of these sediments have to have a more located origin in the surroundings of the check dam. Presumably, the area of the riverbed between a check dam and the next one upstream, in few words, part of the sediments in the check dams are coming from the erosion produced in the riverbed upstream in each check dam.

b. Erosion within the Riverbeds

Precisely, the erosion in the riverbed downstream of the check dams is, in all the stretches studied, quite important and in some cases represents the reason for the breaking of the check dam itself due to the wedging off the check dam's concrete base as it can be appreciated in the example figure 12, where the incision reaches already 65 centimetres of depth and threatens the stability of the check dam's base.

In table 5 we collected the average length of the eroded stretches in each gully that varies from 100 to 200 meters, the average section between 1 and 1.5 square meters, the cubic metres of sediments evacuated in the modified stretches of each gully and its percentage with respect to the total retained sediments in the check dam.

Table 4. Features of the gullies studied and accumulated sediments as a whole in the check dams of each gully

Gullies	Size of basin (hectares)	Number of check dams	Operacional years	Years to silting	Sediments accumulated (m³)
P. del Nevado	1860	17	41	6	23847.4
El Bayo	427	5	41	20	3818.7
Casa del Aire	50	7	41	17	2634.9
Los Pinos	53	4	41	15	2322.0
Los Romanos	593	19	41	30	30557.9
La Pilica	59	5	41	14	5959.3
Total	3042	57	41	17	69140.2

Table 5. Average parameters of the eroded stretches, volume of the eroded soil in all the stretches of each gully and percentage in respect to the total retained in the check dams

Gullies	Average length of eroded (m)	Average section (m²)	Evacuated sediments (m³)	Percentage in respect to the retained soil
P. del Nevado	152	1.5	3876	16.3
El Bayo	135	1	675	17.7
Casa del Aire	105	1.4	1029	39.1
Los Pinos	130	1.2	624	26.9
Los Romanos	198	1.6	6019.2	19.7
La Pilica	115	1.3	745.5	12.5
Total /media	139.2	1.3	10576.7	15.3

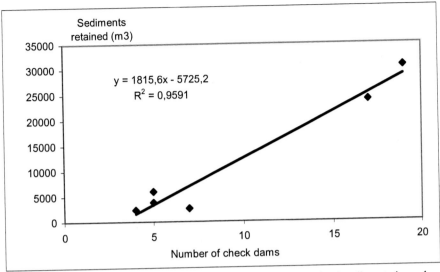

Figure 11. Relation between the number of check dams and the retained sediments in each gully.

Figure 12. Incision within the riverbed downstream in a check dam located in the Los Postes del Nevado gully threatening the stability of the check dam's base.

The contribution of the evacuated sediments in the modified stretches downstream of the check dams to the volume of retained sediments is quite important. In the case of the "La Casa del Aire gully", the sediments supplied by the erosion downstream reaches almost 40 per cent of the total sediments deposited in the check dams. In the "Los Pinos gully" they represent a 27 per cent and in the remaining ones the fluctuations go from 12 to 20 per cent. The average for the whole of the gullies studied is 15 per cent.

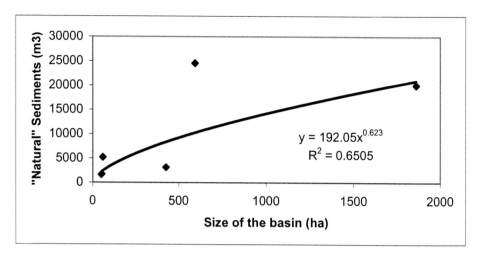

$$y = 192.05x^{0.623}$$
$$R^2 = 0.6505$$

Figure 13. Relation between the size of the basin and the "natural" sediments retained in the check dams.

This represents that in the evaluation of the natural erosion in the gullies affected by hydrological correction works, using as indicators the retained sediments in the check dams, have to be taking into consideration the sediments evacuated from the check dams' riverbed downstream generated by the breaking of the slope due to the construction works.

In fact, if we discount this supply of supplementary sediments (non natural), the relation between the remaining sediments retained in the check dams and the size of the basin, it is very much significant (figure 13).

Obviously, the sediments (non-provoked by erosion of the modified stretches), accumulated in the check dams, have no relation with the size of the basin. Other parameters, such as the slope, lithology or the kind and percentage of the vegetation cover, are basic in this contribution.

4.2. Environmental Effects due to the Opening of Paths

a. The Paths

Almost all the paths studied have been opened starting from agricultural paths next to trails, the access strectches to each check dam are usually short, between 200 and 300 metres, although in some cases, these stretches go over 1 km. In some occasions, the access has been done through crop fields, so the path has been wiped out after the construction of the check dam. In other cases, it has been impossible to determine the place through which the access to the construction of the check dam has been done, probably because it was possible to reach the point without opening path and the forty years passed have wiped off the tracks.

The total length of the opened paths for the check dams' construction is of approximately 10 km. The average width of these paths is of 4 metres and the average depth of the removed soil is 30 cm. With all these, the ploughed area is almost 4 hectares (table 6), representing a volume of soil removed of approximately 12.000 cubic meters, equivalent to 14.300 tonnes.

Table 6. Length, area and volume of removed soil for paths opening giving access to the check dams

Gullies	Path length (m)	Ploughed area (m2)	Removed soil (m3)	Removed soil (Tn)
1	2350	9400	2820	3384
2	1980	7920	2376	2851.2
3	1080	4320	1296	1555.2
4	500	2000	600	720
5	3200	12800	3840	4608
6	825	3300	990	1188
Total	9935	39740	11922	14306.4

Table 7. Length of the path, ploughted area, removed soil and removed soil percentage in respect to the soil retained by the check dams

Gullies	Length of the path per check dam (m)	Ploughted area per check dam (m^2)	Removed soil per check dam (m^3)	Removed soil sediments per check dam (%)
1	138.2	552.9	165.9	6.8
2	396	1584	475.2	35.7
3	154.3	617.1	185.1	28.2
4	125	500	150	14.8
5	168.4	673.7	202.1	7.2
6	165	660	198	9.5
Average	160.2	697.2	209.2	9.9

b. Environmental Cost

The fundamental aim of this research has been the valuing of some environmental effects caused by the opening of paths for giving access to the chosen sites for constructing the sediments retention check dams and, more precisely, the length, area and volume of removed soil by these paths.

The result of this analysis suggest that it will be indispensable to make a previous valuation of the environmental cost provoke by these hydrological correction works, at least, in those places that, because of their difficulties in access, will require the construction of an access path.

The average length of the opened paths for giving access to the construction of check dams in the selected gullies is of approximately 160 meters per check dam, that is equivalent to a plough area of about 700 square metres (table 7) and, in many cases, this area was cover by Mediterranean bushes including protected species by the Comunidad Autonoma de la Region de Murcia (Law Order from February 1989), such as the *Pistacia lentiscus* (Lentisco), *Juniperus oxycedrus, Juniperus comunis* (Enebros), *Quercus coccifera* (coscoja), *Rhamnus lycioides* (Espino negro), *Olea europea Subsp. Sylvestris* (Acebuche) *Tamaris boveana* (Taray) among many others.

Although this average ploughed area per check dam built, is quite homogenous in five of the six gullies, in the gully number 2 ("El Bayo") it reaches 1600 square meters, having the average length of the path for every check dam settled in approximately 400 metres.

For the whole group of gullies studied, the average ploughed area has represented moving approximately 210 cubic meters of soil per check dam built, what it represents approximately 10 per cent of the average volume of the sediments accumulated in the check dams.

From the gullies studied, gully number 2 is especially significant; in it the environmental effects produced by the opening of the paths, are gaining very important dimensions as the volume of the removed soil per check dam, is higher in one third of the volume of retained soil in these ones. This case, with the one in gully number 3 ("Casa Del Aire") a 28 per cent of removed soil in respect to the one retained in the check dams and the gully number 4 ("Barranco de los Pinos") with 15 per cent are especially serious.

5. CONCLUSION

The volume of accumulated sediments along the 41 years of life in the 57 check dams studied represents 70.000 cubic meters, even though, the check dams have not been silted, and the calculations are that they will expand their lifetime between 6 and 30 years.

In a first analysis, the volume of the accumulated sediments in the check dams of each basin is directly related to the number of check dams built in each of them and it is not visible a significant relation between those sediments and the size of the basin. That is why, the accumulation of sediments in the check dams is not valid to us as an indicator of the erosion rate in those basins.

Downstream of the check dams, we can appreciate an incision in all the riverbeds studied as a consequence of the breaking of the slope, provoking within the riverbed the establishing of several levels of local base in each gully.

The average length of the incised or eroded stretches is 140 metres from the base of each check dam, with a length interval between 100 and 200 metres, the average section in the same stretches is of 1.3 square meters.

From the whole group of gullies studied, the evacuated sediments volume in these eroded stretches reaches 10.600 cubic meters, and because the sediments accumulated in the riverbed are deposited in the sediments wedge of the check dam located downstream, this means that 15 per cent of the retained sediments by the check dams come from the erosion in the riverbeds provoked by the construction of check dams itself.

In four of the six gullies, the percentage varies between 12 and 18 per cent, but in the "Los Pinos Gully" it reaches 27 per cent and in the "Casa del Aire gully" represents 39 per cent. This means that a great part of the check dam's capacity must be used in retaining the erosion provoked by the construction of these transversal works in the riverbeds.

Getting rid of the supplying done by the "non natural" supplementary sediments, the relation between the check dams retained sediments volume and the basin's size appears to start becoming relevant. That is to say, the use of the retained sediments volume in the check dams as an indicator of the natural erosion in the basins affected by reforesting hydrological restoration has to take into account the sediments percentage provoked by the erosion in the riverbed due to the check dams construction.

The opening of the access paths to the selected points for the construction of sediments retention check dams becomes indispensable when the difficulties of access are important, but the environmental consequences could be too serious for the little results that in some occasions the check dams have.

The total length of the opened paths in the six gullies studied is approaching to 10 kilometres. This represents a ploughted area of more than 4 hectares and removing 12.000 cubic meters of soil.

These figures show that an average of 160 metres per check dam have been needed for opening a path , this represents a ploughted area of 700 square metres, although, in some cases, the necessary path has been of up to 400 metres and the ploughted area of approximately 1600 square meters.

All of these represent an environmental cost that would be necessary to consider before planning any new opening of paths, because the removed soil average volume for the opening of these paths reaches 10 per cent of the retained soil volume in the check dams 40 years after their construction, even reaching almost 36 per cent in the gully number 2, 28 per cent in the gully number 3 and 15 per cent in the gully number 4.

With all these, the worst is that these paths were abandoned, after their construction, at the erosive process will, without carrying out any vegetation restitution work, and even less, the elimination of the path itself.

6. ACKNOWLEDGEMENTS

This work has been done within the framework of the research project reference REN2002-03426/HID, funded by the Ministry of Science and Technology and FEDER Funds. To all of them, our most sincere recognition.

7. REFERENCES

Avendaño Salas, C., Sanz Montero, E., Cobo Rayan, R. and Gómez Montaña, J.L. (1997). Sediment yield at Spanish reservoirs and its relationship with the drainage basin area. *Dix-neuvième Congrès des Grands Barrages,* 863-874. Commsission Internationale des Grands Barrages. Florence.

Alonso Sarría, F., Belmonte Serrato, F., Marín Sanleandro, P., Martínez Lloris, M. Ortíz Silla, R., Rodríguez Estrella, T., Romero Díaz, A. and Sanchez Toríbio, M.I. (2003): La función de los diques de retención de sedimentos y su relación con las características litológicas de sus cuencas. Cuenca del río Quipar, Región de Murcia. In R. Bienes y M.J. Marques (Eds.) *Perspectivas de la degradación del suelo.* Instituto Madrileño de Investigación Agraria y Alimentaria. Madrid.

Brandt, S.A. (2000): Classification of geomorphological effects downstream o dams. *Catena,* 40: 375-401.

Conesa García, C., Belmonte Serrato, F. and García Lorenzo, R. (2004): Efectos de los diques de corrección hidrológico-forestal en la competencia y estabilidad de corrientes efímeras. Aplicación a la cuenca de la rambla de la Torrecilla (Murcia) In *Territorio y medio*

Ambiente: Métodos cuantitativos y Técnicas de Información Geográfica. C. Conesa García y J.B. Martínez Guevara (eds.). 69-83. Murcia.

CHS (1962). Proyecto de Corrección Hidrológica para la protección del Embalse de Alfonso XIII. Ministerio de Obras Públicas. Dirección General de Obras Hidráulicas. Confederación Hidrográfica del Segura.

CHS (1992). Proyecto de Corrección Hidrológica de las cuencas de Alfonso XIII y del Cárcavo. Ministerio de Obras Públicas y Transportes.

García Ruiz, J.M. and Puigdefábregas, J. (1985): Efectos de la construcción de pequeñas presas en cauces anastomosados del Pirineo Central. *Cuadernos de Investigación Greográfica.* 11: 91-102.

Hernández Laguna, E., Martínez Lloris, M. and Romero Díaz, A. (2004): Determinación del volumen de sedimentos retenidos en diques de corrección hidrológica. *VIII Reunión Nacional de Geomorfología.* Toledo. In G. Benito and A. Díez Herrero (Eds.), *Riesgos Naturales y Antrópicos, en Geomorfología.* SEG y CSIC, Voll, 201-210.

Martínez de Azagra, A., Fernández de Villara, R., Seseña Rengel, A., Méndez Carvajal, C., Díez Hernández, J.M., Navarro Hevia, J. and Varela Nieto, J.M. (2002). Metodología para la inventariación de diques forestales gavionados. Aplicación en la provincia de Palencia. *Cuadernos de la Sociedad Española de Ciencia Forestal,* 13: 171-181.

Martínez Castroviejo, R., Invar, M., Gómez-Villar, A. and García Ruíz, J.M. (1990): Cambios en el cauce aguas debajo de una presa de retención de sedimentos. In *Actas de la I Reunión Nacional de Geomorfología.* 457-468. Teruel.

Martínez Lloris, M., Romero Díaz, A. and Alonso Sarría, F. (2001): Respuesta erosiva de cuencas, corregidas mediante diques de retención de sedimentos, ante lluvias de alta intensidad. Cuenca del río Quipar, Sureste de España. *Papeles de Geografía,* 34: 191-203.

Ministerio de Medio Ambiente (MMA) (2006). *Restauración hidrológico forestal. Conservación de la Naturaleza, Acciones.* En: http://www.mma.es/conserv_nat/acciones-/restaur_hidro/rest_hidro.htm.

Romero Díaz, A., Cabezas, F. and López Bermúdez, F. (1992). Erosion and fluvial sedimentation in the river Segura basin (Spain). *Catena,* 19: 379-392.

Romero Díaz, A., Martínez Lloris, M. and Belmonte Serrato, F. (2004): The construction of dikes of hydrological correction as policy to retain the erosion and to avoid the silting up of dams in the Segura basin (Spain). *Fourth International Conference on Land Degradation.* In A. Faz, R, Ortiz and G. García (Eds).

Romero Díaz, A., Martínez LLoris, M., Alonso Sarría, F., Belmonte Serrato,F., Marín Sanleandro, P., Ortíz Silla, R., Rodríguez Estrella, T. and Sánchez Toribio, M.I. (2007). *Los diques de corrección hidrológica. Cuenca del Río Quipar (Sureste de España).* Ediciones de la Universidad de Murcia, 270 pp., Murcia.

Sánz Montero M.E., Avendaño, C., Cobo, R. and Gómez, J.L. (1998). Determinación de la erosión en la Cuenca del Segura a partir de los sedimentos acumulados en sus embalses. *Geogaceta* 23: 135-138.

Valera Nieto, J.M. (Dir) (1999). Inventario de obras de corrección hidrológica-forestal y de las variables ambientales relacionadas con las mismas. Centro de Estudios de Técnicas Aplicadas. Dirección General de Obras Hidráulicas.

World Commision on Dams (WCD) (2000). *Dams and development: A new framework for decision-making.* Earthscan Publication Ltd., London.

In: Dams: Impacts, Stability and Design
Editors: Walter P. Hayes and Michael C. Barnes

ISBN 978-1-60692-618-5
© 2009 Nova Science Publishers, Inc.

Chapter 7

Changes in a River-Coastal Aquifer System and Saltwater Intrusion Risk Related with Recent Dam Construction (Motril-Salobreña Aquifer, S Spain)

Carlos Duque[1]*, Wenceslao Martín-Rosales[1],
María Luisa Calvache[1], Manuel López-Chicano[1],
Antonio González-Ramón[2] and Juan Carlos Rubio Campos[2]

1. Departamento de Geodinámica. Facultad de Ciencias, Universidad de Granada. Avda. Fuentenueva s/n, 18071 – Granada
2. Instituto Geológico y Minero de España, Oficina de Proyectos de Granada, Urb. Alcázar del Genil, 4-Edif. Zulema, Bajo, 18006 - Granada

Abstract

Southern Spain has increasing water-supply problems due to the expansion of cities, tourism, and farming areas. In 2005, the Rules Dam was constructed on the Guadalfeo River 20 kilometres from the sea in the search for improved water provisions for urban and irrigation uses. Subsequently, river discharge decreased, especially in the sector nearest the river mouth, where the river is now permanently dry. For centuries, the discharge of this river and irrigation excess (previously derived from the river) were the main sources of recharge for the Motril-Salobreña coastal aquifer, formed by the deposition of sediments at the mouth of the Guadalfeo River. The interruption to river flow is likely to decrease these inputs to the aquifer, most probably causing a reduction in aquifer resources that would then affect the behaviour of the hydrogeological system. Therefore, the study of the influence of these factors on the water table could provide crucial information on the dam impacts on groundwater. One of the main secondary effects due to a water-table decrease in coastal aquifers is saltwater intrusion. Along the neighboring coastal aquifers, seawater encroachment has been detected, but in the Motril-Salobreña aquifer, the influence of the river maintained (until recently) optimal quality

* E-mail: cduque@ugr.es.

and quantity of groundwater. The study of marine intrusion is very important due to possible changes; the main problem is the great aquifer thickness (more than 200 metres) and the lack of boreholes for the measurement of groundwater conductivity. The combination of two geophysical techniques has been very useful for the quantification of the state of this aquifer. The monitoring of the possible effects from the decrease in the Guadalfeo River flow, the water-table drop and, finally, saltwater encroachment, as the main hydrogeological consequences of the Rules Dam, may allow the dam's negative impacts to be minimised by planning for sustainable management.

INTRODUCTION

Dam construction in semi-arid zones has been aimed at a better distribution of water resources. Mediterranean areas are characterised by a climate in which most of the annual precipitation falls on just a few days, which makes such engineering works essential to guaranteeing continuity in supplies. Therefore, the role of such dams is twofold: to store water during rainy periods and to store then gradually distribute the storm runoffs that periodically occur in these areas. Construction of a dam tends to bring richness to zones where lack of water has limited economic development. It must not be forgotten, however, that such engineering works can have other impacts.

The effect of dams on river flow is well known (Shalash, 1983; Benjamin and Van Kirk,1999; Ma et al. 2007; Milliman et al. 2008), but a reduction in discharge may not be the only hydrogeological change in affected zones. It must be kept in mind that surface and ground waters are not independent and that effects on a river may have repercussions on an aquifer's behavior. Therefore, when constructing a dam, it behooves us to take into account not only the impact on the rivers in the area, but also on the aquifers recharging from those rivers.

In the case of the Rules Dam (S Spain), the Guadalfeo River plays a crucial role in the hydrogeological system of the Motril-Salobreña aquifer just fifteen kilometres downstream from the dam (Figure 1). The high permeability of the riverbed results in very high water infiltration, making it a significant source of recharge for the aquifer. Since the Rules Dam intercepts the Guadalfeo River, the reduced volume of water downstream undeniably affects the phreatic levels; in turn, the drop in the phreatic levels leads to deepening of wells and the risk of over-abstraction from the aquifer. Moreover, as this is a coastal aquifer, one of the greatest dangers of dropping phreatic levels is marine intrusion. The relations among the different types of effects on the hydrogeology in the area surrounding the Rules Dam are analysed in this work to establish what drawbacks have arisen from its construction.

GEOGRAPHICAL AND HYDROGEOLOGICAL SETTING

The Motril-Salobreña detrital aquifer (SE Spain) is surrounded by high reliefs on all sides except the south, where it meets the Mediterranean Sea. To the northeast is the Sierra de Lújar, a range reaching over 2,000 metres elevation less than 20 kilometres from the coastline. To the northwest are the Sierra de la Almijara and Sierra de Tejeda, also peaking at over 2,000 metres (Figure 1). These ranges indirectly nourish the aquifer through their runoff into the Guadalfeo River, although most of the volume derives from the peaks of Sierra Nevada, which rises to over 3,000 metres elevation (e.g. Mulhacén 3,482 m.a.s.l.; Veleta: 3,392 m.a.s.l.; Alcazaba: 3,364 m.a.s.l.).

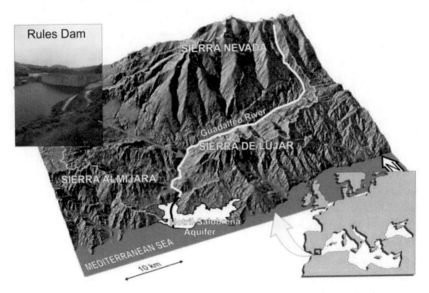

Figure 1. Geographic location of the Motril-Salobreña aquifer and its catchment basin.

Figure 2. Catchment basin of the Guadalfeo River and drainage system showing the main rivers.

The aquifer's name derives from the two main towns sited above it. Both Motril (about 60,000 inhabitants) and Salobreña (over 10,000 inhabitants) massively increase in population during the summer due to the influx of tourists.

The Guadalfeo River is fed by water from various streams originating in Sierra Nevada, such as the Trevélez, the Poqueira, and the Lanjarón. Also of importance is the Ízbor River (Figure 2), running through the western zone of the range and whose main tributary is the Dúrcal River. Since all of these water courses begin at high elevations, snowmelt plays an important role in the volume of the Guadalfeo at the end of spring and start of summer.

The climate in the study sector is mainly controlled by its proximity to the sea and by its latitude. It is a temperate Mediterranean climate (locally the area is known as the tropical coast) with considerable cultivation of subtropical plants such as avocado, cherimoya, and mango. Temperatures are very mild throughout the year (annual average of 18 °C) with minimal variations compared to the extremes in nearby cities that are farther from the coast. Precipitation is scarce (annual average of 435 mm in Motril and 416 mm in Salobreña). Precipitation and temperature vary considerably throughout the Guadalfeo's catchment basin, and since it influences the recharge of the Motril-Salobreña aquifer, we will examine the precipitation across the entire area drained by the river (Figure 2).

The isohyetal map (Figure 3) shows the low average annual precipitation of the coastal belt and how the increase at higher elevations in the northern sector of the basin. There is a clear gradient linking topographic elevations at each point with the amount of annual rainfall. This pattern is evident in the ranges of Almijara and Tejeda northwest of the aquifer and is even clearer in the Sierra Nevada due to its higher elevation. The cloud fronts driving in from the west must rise over these mountainous reliefs and therefore drop their moisture content as elevation increases, with the highest amounts (1,000 mm/year) falling on the peaks of Sierra Nevada and the least at the coast, where there are no geographic highs to force precipitation.

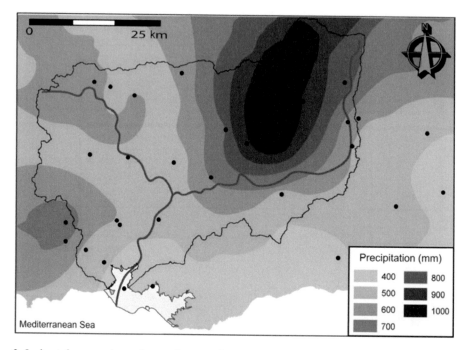

Figure 3. Isohyetal map and weather stations used to plot the isohyetal lines.

The study area has a complex tectonic history, with varying degrees of deformation, metamorphism, and fracturing affecting most of the materials forming the reliefs of the catchment basin (García-Dueñas and Avidad, 1972), with rocks dating back as far as the Permian-Triassic (Aldaya, 1981). The erosion of these mountain systems has produced a complex drainage system, with the formation of very recent Quaternary alluvial systems in some zones.

Materials can be classified simplistically by their hydrogeological properties as units with scarce permeability (schists and phyllites) and those with high permeability (especially carbonates and Quaternary sediments) (Figure 4). The generalised scarce permeability of this zone, together with other characteristics such as steep slopes and storm-like precipitation, favors surface runoff.

Quaternary sediments form good detrital aquifers (e.g. Motril-Salobreña) and carbonates form good karst aquifers. In both cases, the main recharge is caused by infiltration deriving from surface runoff and direct rainfall.

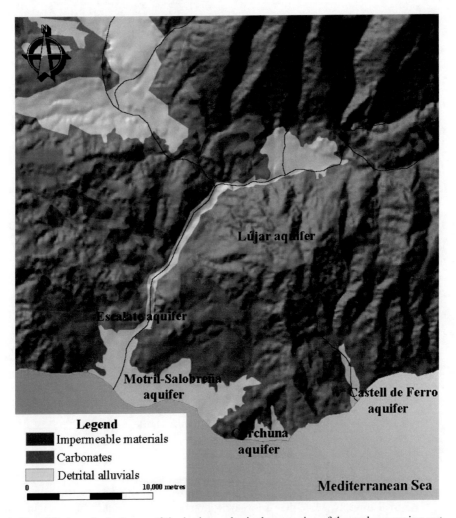

Figure 4. Simplified geological map of the hydrogeological properties of the rocks cropping out.

The Motril-Salobreña aquifer is formed by alluvial sediments transported by the Guadalfeo River and its tributaries and, to a lesser extent, by a few other small dry washes. This aquifer is the most important detrital coastal aquifer in the province of Granada, due both to its size and its water resources (Castillo and Fernández-Rubio, 1978; Pulido-Bosch and Rubio, 1988; ITGE, 1991; Calvache et al., 2003). Moreover, the water quality has been very high for at least the last 30 years. In its western sector, the aquifer lies beneath the course of the Guadalfeo River, which has a mixed rainfall-snowfall water regime. The river undergoes a net loss of water for most of its course until it reaches its distalmost section, where the aquifer discharges into the river. Input to the aquifer from direct precipitation is scarce due to the Mediterranean climate, in which precipitation is scarce and evapotranspiration very high. The main recharge is from infiltration from the bed of the Guadalfeo River and from excess irrigation water (Ibáñez, 2005), which was traditionally obtained from diverting river water (Figure 5).

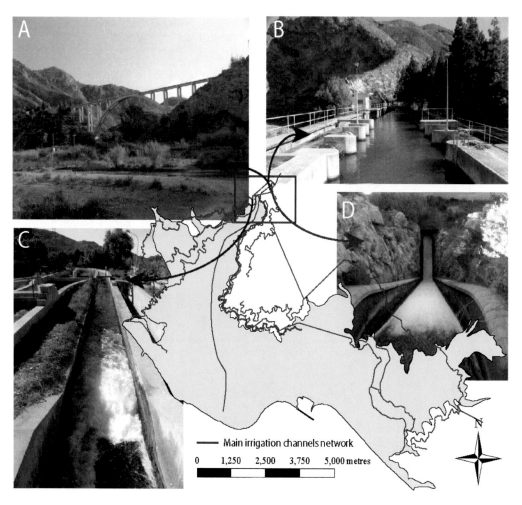

Figure 5. Water diversion from the Guadalfeo River. A. Cañizares aqueduct. B. Cañizares distributor (hydraulic works to distribute water to the various irrigation channels). C. One of the main irrigation channels in the Vínculo irrigation dam. D. The water is transported down great drops in elevation by this system.

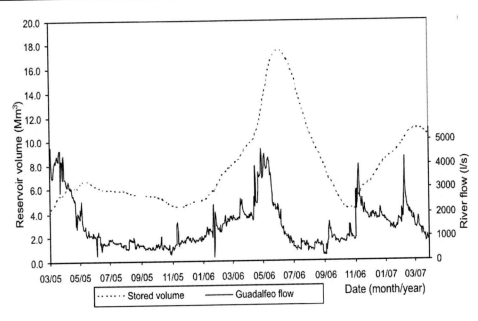

Figure 6. Evolutions of volume of water stored in the reservoir and the volume of the Guadalfeo River.

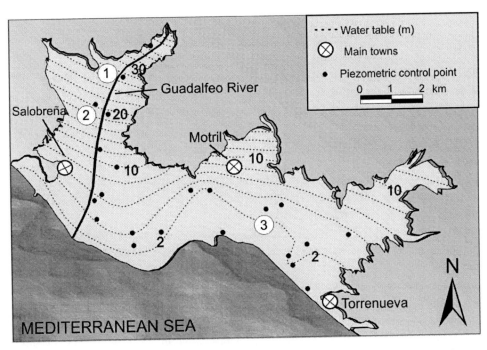

Figure 7. A water-table map for the Motril-Salobreña aquifer showing also various control points used for a more detailed analysis.

Lithological columns show that the gravel, sands, silts, and clays forming the aquifer vary in proportion in different sectors due to the sedimentary conditions reigning during the formation of the aquifer (Duque et al., 2005). This circumstance is reflected in the fact that the hydrogeological properties of this detrital aquifer, the hydraulic conductivity, and the specific yield vary considerably by sector. The aquifer is nearly entirely surrounded by impermeable materials except to the north, where there are carbonates (Escalate aquifer) hydraulically connected with it (Figure 4), and the alluvials of the Guadalfeo itself, also connected with the Motril-Salobreña aquifer. As irrigation needs are mainly supplied by surface water, pumping is not intensive, being mostly employed for the scarce industry in the area and for urban water supply of towns, primarily Motril and Salobreña. The principal discharge of water from the system is by hidden discharge to the sea, thus impeding advance of the saline wedge (Duque et al., 2008).

Various other coastal detrital aquifers in this zone have also been studied (e.g. Carchuna, Castell de Ferro, and Río Verde aquifers), although they are much smaller than Motril-Salobreña and much less hydrogeologically complex. These aquifers have particularly been examined in relation to marine intrusion (Calvache and Pulido, 1991, 1994, 1997) since saline-wedge advancement has been common, particularly during droughts. There is no doubt that the lack of a river like the Guadalfeo and high abstraction rates by humans have facilitated the advance of the saline wedge in those aquifers.

RULES DAM

Rules Dam was constructed to hold back the waters of the Guadalfeo River, which is particularly voluminous given the environment it runs through due to its huge catchment basin (1,200 km^2), including the high peaks of Sierra Nevada. The geology of the catchment basin plays an important role in the river's hydraulic behaviour since the predominance of impermeable materials favours surface runoff and increases surface flow to the detriment of a heavier recharge of the aquifers. The slopes of the mountain ranges are steep (due in large part to their geological youth), which strongly determines the circulation of surface waters and the drainage patterns.

The dam is an arch-gravity dam of conventional vibrated concrete, a radius of 500 metres, and a height of 130 metres. Its outflow capacity is 6,100 m^3/s, with a fixed-lip spillway of 140 metres wide. Construction required the extraction of 2,600,000 m^3 since the dam intercepts both the surface and the alluvium of the Guadalfeo River. The dam also required 2,200,000 m^3 of concrete (MMA, 2005).

The dam was built both for urban water supply and irrigation since droughts are quite common in this zone and summers are very long and dry. The coast of Granada has been undergoing rapid development in recent years due to changes in techniques for traditional farming and to an increase in tourism driven by the mild climate reigning for most of the year. Consequently, the number of year-round residents of towns such as Motril has nearly doubled in the last 35 years to reach its current peak of 60,000 permanent inhabitants with a corresponding increase in water consumption. Urban planning is counting on expansion of the town and an increase in tourism, as noted by construction projects of residential zones with golf courses.

The traditional crop grown on the plain of Motril-Salobreña until very recently was sugar cane, introduced by the Arabs hundreds of years ago. Having at their disposal the huge richness of water supplied by the Guadalfeo River, those inhabitants designed and built a great number of small hydraulic works to distribute water by gravity from the river to all the croplands of the plain (Figure 5). Therefore, the irrigation system was flood-type, which transformed the plain into a zone where the croplands were a significant source of aquifer recharge since some of the irrigation water infiltrated back into the ground. Over more recent years, sugar cane has given way to more economically attractive crops such as subtropical fruit trees or extra-early greenhouse crops, which are already the dominant types of crops in the nearby province of Almería (Parra et al, 2008; Van Cauwenbergh, 2008). As a result of these changes, the irrigation systems have evolved, and flood irrigation has been replaced in many cases by drip irrigation, which uses much less water but provides almost no return flow. Hothouse agriculture, which is very economical, has allowed slopes several hundred metres high to be cultivated and irrigated with water pumped from lower zones. In short, the irrigation requirements have dropped, but the land area under cultivation has increased, so water consumption has remained about the same. The greatest difference is that irrigation return flow has been significantly reduced, resulting in less recharge for the aquifer.

The direct goal of the dam, therefore, is to obtain better water management of the resources of the Guadalfeo River, although it is affecting management of the Motril-Salobreña aquifer due to the close relationship between the two.

IMPACTS OF THE RULES DAM

Affect on Volume of Flow

Dam filling began March 4, 2005, reaching its maximum volume to date (March 31, 2007) on June 5, 2006, with 17.5 Mm^3 of water in the reservoir pool. Since first filling began up to March 31, 2007, there have been three peaks in inflow (see Figure 6). As expected, these episodes coincide with moments when the river volume is high enough.

The largest took place in 2005-2006 and lasted nearly 200 days, resulting in the net storage of over 13 Mm^3.

The average volume of flow measured in the bottom outlet has been nearly constant during the period for which data are available. The maximum outflow by the dam was 3,875 l/s (on April 1, 2005), at which time the incoming flow was 4,401 l/s; the minimum volume released (January 17, 2006) was 250 l/s, with an incoming flow of 1,280 l/s. The average outflow volume is 1,595 l/s, and the average input flow is 1,730 l/s. Consequently, the operation of the dam has meant a reduction in the river volume of flow of approximately 8% (to March 31, 2007).

These outflows are mainly diverted a few kilometres downstream at the Vélez weir for urban supplies and irrigation. Therefore, downstream of this weir at the Vínculo weir (Figure 5), one of the northwestern boundaries of the detrital aquifer, the volume of flow is lower still. Several campaigns on flow volume carried out from 2001 to 2004 measured over 7,000 l/s of volume of flow in the river near the mouth. Since the dam entered operation, however, the river flow downstream of Vínculo weir is almost non-existent.

Although quantitatively the degree to which the river is affected does not seem overly high, observations in the study zone confirm that, since the construction of the Rules Dam, the flow of the Guadalfeo above the Motril-Salobreña aquifer has been almost nil. In addition, what flow there is seems to be related with supplies by small streams downstream of the dam.

This is a clear indication there has been scarcely any recharge of the aquifer by the river and that the decrease in the river's volume of flow is most drastic in those reaches where the river's course takes it over the surface above the coastal aquifer.

Effect on the Water Table

Various studies carried out over the last 30 years have shown great stability in the water table. The water-table map for the Motril-Salobreña aquifer (Figure 7) shows a greater hydraulic gradient in the northern sector, with a gradual decrease in gradient towards the coast (González et al., 2005). Note also that the gradients are sharper in the western section than the eastern one, most likely due to the lower recharge in this zone from the Guadalfeo River (Calvache et al., 2004).

Three piezometric control points were chosen from the network to determine the main effects on the aquifer. Two of these control points are in the northern sector very close to the Guadalfeo River and the third is several kilometres away from the river and nearer the coastline (Figure 7).

The plot for piezometric evolutions (Figure 8 and 9) shows that levels were relatively stable up until March 2005, with control point 2 recording levels of 20.2 and 26.7 m and control point 1 26-36 m. At control point 3, farther east, the level is closer to the surface and farther from the river, so effects from the river are less evident. The evolution of the level is quite consistent with the dynamics of precipitation recorded in Motril (the interannual average for 1954/55-2005/06 is 410 mm) up until at least 2005, at which point the decrease is practically continuous, with just a few minor recoveries. In 2001/02 it rained 348 mm and the water-table levels rose about 1.4 m in the western sector. In 2002/03 and 2003/04, precipitation was 493 mm and 481.4 mm, respectively, with rises of 9 m in the water-table levels in the western sector, but rises of only about 2 m in the eastern sector.

Since 2005, however, there has been a general drop in the water-table level in all parts of the aquifer, although slightly more in the western sector near the river. From the time the dam began operations to January of 2006, piezometric falls have ranged from 3.5 m at point 2 to 6 m at point 1 in the western sector, but only about 1 m in the eastern sector. The advent of rainfall periods produces a much lower recovery than previously, generating smoother recharge peaks quite differently than in the two years prior to dam operations and always in a clear downward trend.

The graphs of the piezometric evolution of diverse sampling points over the last few years indicate a general drop in levels. Two parameters affect this drop—distance from the river makes the effects less visible since points nearest the river were most affected by its recharge; in addition, distance to the sea also influences the range in levels as changes are greater in northern sectors of the aquifer and much smaller near the sea.

The variations over the last few years show much more abrupt changes since the Rules Dam entered into operation. The falls are very similar to those that occurred prior to the dam, but the recoveries that used to happen every year are now very weak. It seems this effect may be related to the presence of the dam, since it is precisely this surface flow that passed above the aquifer and provided a significant percentage of its recharge and has now been practically eliminated to optimize the water resources of the Guadalfeo River.

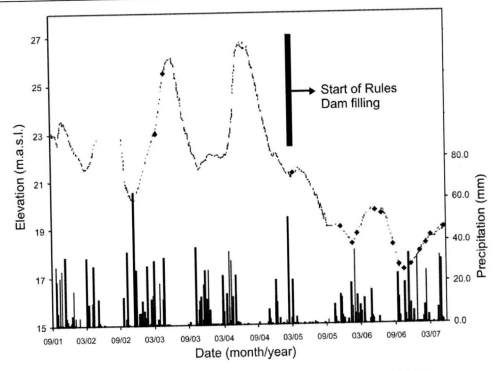

Figure 8. Piezometric evolution of control point 2 and daily precipitation for 2001-2007.

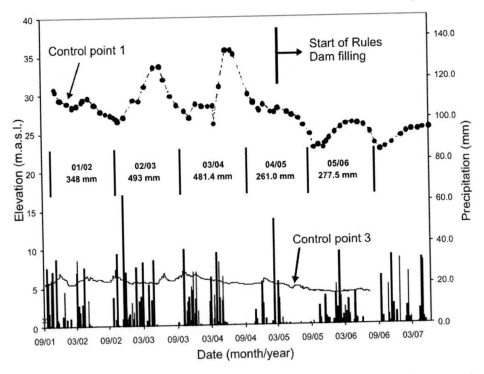

Figure 9. Piezometric evolution at control points 1 and 3, daily precipitation at a weather station above the Motril-Salobreña aquifer, and cumulative annual precipitation.

Nevertheless, these results must be taken cautiously since the last few years have been dry ones within the predominant climate in the zone (see Figure 9). Consequently, the effects observed in Figures 8 and 9 probably are not solely the result of the direct influence of the Rules Dam. Rather, they are the result of the sum of the effects of the dam and the lack of precipitation during the same period that have combined to cause the drop in aquifer levels.

This convergence should be studied over the next few years to separate the influence of each of these two factors and thereby reach quantitative estimations of the consequences of the regulation of the surface flow by the Rules Dam.

Monitoring of Saltwater Intrusion

Marine intrusion is a common risk in coastal aquifers. The contact point of seawater with freshwater remains stable as long as the hydraulic conditions of the aquifer are also stable. One of the worst consequences of aquifer salinization is its irreversibility (at least in the short and middle term) of the consequences on the groundwater (Custodio y Llamas, 1983; Custodio, 2005) and indirectly on the environment due to the use of water with high salt contents. Therefore, research on marine intrusion focusses on preventing its encroachment as much as possible given the fact that aquifers saturated with brackish waters require extremely long periods of time to regain water quality. Moreover, using water with high salt contents is very damaging to plant growth and sterilizes the soil (Magán et al., 2008; Manchanda and Garg, 2008; Yermiyahu et al., 2008), thereby inhibiting both crop growth and autochthonous vegetation. Since the level of the Motril-Salobreña aquifer has been dropping over the last few years, attempts must be made to detect the effect this is having on the saline wedge.

The usual means of locating the saline wedge along the coastline by measuring the conductivity of the groundwater is not always possible due to a lack of sufficient wells. Well-drilling is extremely expensive and, since the aims are purely scientific (as reaching brackish water is not generally useful), financing is very difficult to find. Consequently, geophysical techniques have traditionally been very popular for monitoring changes in the position of the saline wedge (Gondwe, 1991; Kafri et al, 1997, Chen, 1999; Hopkins and Richardson, 1999; Choudhury et al., 2001; Guérin et al, 2001; Balia et al. 2003; Goldman et al., 2003; Wilson et al., 2005). Studies along coastlines have commonly made use of Vertical Electrical Soundings (VES) and, more recently, of Time Domain Electromagnetic Soundings (TDEM) due to the electrical properties of saltwater and the changes it causes in aquifer materials.

Interpretation of the results provided by these geophysical techniques is often complicated by geological complexity and by the fact that a given resistivity may be due to different materials with distinct degrees of saturation or to variations in water conductivity. Progress continues in this area as technical improvements are made in the measuring systems and geophysical data are complemented with other data to successfully resolve the uncertainties inherent in these methods.

The Rules Dam has affected the volume of the Guadalfeo River downstream of it and water for human and agricultural use is being used more efficiently, leading to a lowered recharge of the Motril-Salobreña coastal aquifer. It is important to locate the saltwater-freshwater interface in order to evaluate changes and the existing relationship with the phreatic levels. It must be kept in mind that this aquifer has an annual input of some 35 Mm^3, so a reduction in input need not necessarily result in a sharp inland advance of the interface.

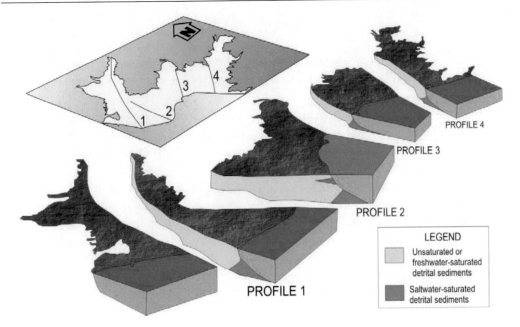

Figure 10. Results from combined geophysical campaigns of TDEM soundings and gravimetry.

In November of 2006, a geophysical campaign was carried out with TDEM soundings to locate the saline wedge along the Mediterranean coastline in four sectors of the Motril-Salobreña aquifer (Figure 10). At that point, the dam had been in operation for very little time, so the results can be considered as pre-dam conditions or, at least, as incipient conditions. Difficulties were found in the initial interpretation due to the location of low resistivities in various zones at different depths that might have been interpreted as saltwater intrusion (Duque et al., 2006). Some of these readings could be related to the low resistivities properties of the materials comprising the aquifer basement (Geirnaert et al., 1981).

In consequence, a gravimetric campaign was planned in order to establish the aquifer depth and thereby have a reference line for determining which materials are marine-water saturated. The technique of combining gravimetry and electromagnetic methods can be used to evaluate the aquifer properties based on two very different parameters, thereby enriching the original data. Gravimetry determines rock density and so allows researchers to differentiate between the poorly consolidated material of the detrital aquifer and the hard rock comprising the basement. Electromagnetic methods determine electromagnetic properties, thereby allowing a differentiation between detrital materials saturated in saltwater because of the drop in measurable resistivity.

The results shown in Figure 10 clearly indicate where marine intrusion has been detected. This situation is that of a state of equilibrium or of one very little modified by changes in aquifer levels. With this information, successive monitoring campaigns will be able to establish changes in levels and in the saltwater intrusion. Moreover, as changes in the hydraulic regime of the Guadalfeo River due to the dam and any effect on aquifer levels can also be evaluated, we will finally be able to correlate the effects of the dam on saltwater intrusion if it should occur in the Motril-Salobreña coastal aquifer.

CONCLUSION

The hydrogeological effect that the construction of a dam may exert on a river's course is not limited just to interrupting the flow of surface water as river water may form part of more complex hydrogeological systems.

In the case of the Guadalfeo River–Motril-Salobreña aquifer, the recent construction of the Rules Dam has reduced the volume of flow in the Guadalfeo by 8% to date. This decrease has particularly affected the amount of water in the Guadalfeo's riverbed in its reach over the Motril-Salobreña coastal detrital aquifer.

Due to the significant role of the Guadalfeo River in recharging the Motril-Salobreña aquifer, water-table records (particularly near the course of the river) appear to be dropping since the dam entered operation. These data, however, must be interpreted with caution due to the difficulty in differentiating between effects caused by periods of dry climate and the direct effects of this work of engineering.

Marine intrusion is the most serious indirect effect that could occur in the Motril-Salobreña aquifer as a result of the drop in recharge produced by water storage in the reservoir. Monitoring saltwater encroachment, using the results from previous geophysical campaigns as a starting point, will allow the determination of the relationship that may exist between the position of the saline wedge and the containment of the flow of the Guadalfeo River.

The interrelationship of the components in hydrogeological systems may make an effect on one rebound on the rest. The improvement in usage of surface waters by dam building should be accompanied by monitoring of the direct and indirect consequences that may derive from their operation in order to undertake a sustainable use of such resources and minimise negative impacts.

ACKNOWLEDGMENTS

This study was made possible by funding approved for projects REN2000-1377HID and CGL2004-02748/HID financed by the Ministerio de Educación y Ciencia of Spain and the Spanish program of Formación de Personal Investigador. We thank Christine Laurin for the English text.

REFERENCES

Aldaya, F. 1981. Hoja nº 1056 (Albuñol del Mapa Geológico de España a escala 1:50000). IGME. Madrid.

Balia R., Gavaudò E., Ardau F., Ghiglieri G. 2003. Geophysical study of a coastal plain. *Geophysics* Volume 68, Issue 5, September 2003 pages 1446-1459

Benjamin, L., Van Kirk, R.W. 1999. Assessing instream flows and reservoir operations on an eastern Idaho River. *Journal of the American Water Resources Association* 35 (4), pp. 899-909

Calvache, M.L., Rubio, J.C., López-Chicano, M., González-Ramón, A., Cerón, J.C., Ibáñez, S., Martín-Rosales, W. Soler, R., Díaz-Losada, E. y Peinado-Parra, T. 2003. Estado actual del acuífero costero de Motril-Salobreña previo a la puesta en funcionamiento de la presa de Rules (Granada, España). In: Proceedings of Tecnología de la Intrusión de Agua de Mar en acuíferos costeros: Países mediterráneos. Alicante, 1: 77-86.

Calvache, M.L., Cerón, J.C.,Rubio, J.C., Martín-Rosales, W, López-Chicano, M., González-Ramón, A., Ibáñez, S., Duque, C. 2004. Caracterización de las relaciones río-acuífero en el sistema Motril-Salobreña (Granada). In: Proceedings of VIII Simposio de Hidrogeología. Zaragoza, 433-442.

Calvache M.L. y Pulido-Bosch A. 1991. Saltwater intrusion into a small coastal aquifer (rio Verde, Almuñécar, S. Spain). *Journal. of Hydrology* 129:95-213

Calvache M.L. y Pulido-Bosch A. 1994. Modelling the effects of salt water intrusion dynamics for coastal karstified block connected to a detrital aquifer. *Ground Water* 32(5):767-777

Calvache M.L. y Pulido-Bosch A. 1997. Effects of geology and human activity on the dynamics of salt-water intrusión in three coastal aquifers in southern Spain. *Environmental Geology* 30:215-223

Castillo, E. 1975. Hidrogeología de la Vega de Motril-Salobreña y sus bordes. Master Thesis. Universidad de Granada. 184 p.

Castillo, E. y Fernández Rubio, R. 1978. Hidrogeología del acuífero de la Vega de Motril-Salobreña. Bol. IGME, LXXXIX, 39-48. Madrid.

Chen C.S. 1999. TEM investigations of aquifers in the Southwest coast of Taiwan. Groundwater 37 (6):890-896 NOV-DEC 1999

Choudhury K., Saha D.K., Chakraborty P. 2001. Geophysical study for saline water intrusion in a coastal alluvian terrain. *Journal of Applied Geophysics* 46, 2001. 189-200

Custodio E., 2005. Coastal aquifers as important natural hydrogeological structures. In Groundwater and Human Development by Emilia M. Bocanegra, Mario Alberto Hernández, Eduardo Usunoff (editors). International Association of Hydrogeologists. *Taylor and Francis* 262 p. ISBN:0415364434

Custodio E. y Llamas M.R., 1983. Hidrología subterránea. Editorial Omega p. 2350.

Duque, C.,Calvache, M.L., Pedrera, A., Martín-Rosales, W., López-Chicano, M. 2008. Combined time domain electromagnetic soundings and gravimetry to determine marine intrusion in a detrital coastal aquifer (Southern Spain). *Journal of Hydrology*, 349,536-547. DOI: 10.1016/j.jhydrol.2007.11.031

Duque, C., Calvache, M.L., Rubio, J.C., López-Chicano, M., González-Ramón, A., Martín-Rosales, W. and Cerón, J.C. 2005. Influencia de las litologías en los procesos de recarga del río Guadalfeo al acuífero de Motril-Salobreña. In: Proceedings of VI Simposio del Agua en Andalucía, 343-355. Sevilla.

García-Dueñas V. and Avidad J. 1972. Hoja nº 1055 (Motril del Mapa Geológico de España a escala 1:50000). IGME. Madrid.

Geirnaert, W., Pulido-Bosch, A., Castillo, E. y Fernández Rubio, R., 1981. Estudio de la geometría del acuífero detrítico de la vega de Motril-Salobreña mediante SEV. In: Proccedings of Simposio del Agua en Andalucía, 291-302.

Goldman M., Kafri U. y Yechieli Y. 2003. Application of the time domain electromagnetic (TDEM) method for studying groundwater salinity in different coastal aquifers of Israel.

In: Proceedings of Simposio Internacional sobre Tecnología de la Intrusión de agua de mar en Acuíferos Costeros. IGME, Madrid, pp. 77-85.

Gondwe, E. 1991. Saline water intrusion in Southeast Tanzania. *Geoexploration* 27 (1-2): 25-34 Feb.

González-Ramón, A., Peinado-Parra, T., Calvache, M.L., Rubio, J.C., López-Chicano, M., Martín-Rosales, W., Navarro, J.A., 2005. Evolución de niveles piezométricos en acuíferos deltaicos. Nuevas aportaciones al conocimiento hidrogeológico del acuífero Motril-Salobreña. En: Proceedings of VI Simposio del Agua en Andalucía. Sevilla, pp. 355-36

Guérin, R., Descloitres, M., Coudrain A., Talbi A., Gallaire R. 2001. Geophysical surveys for identifiying saline groundwater in the semi-arid region of the central Altiplano, Bolivia. *Hydrological Processes*. 15, 3287-3301

Hopkins D.G. and Richardson J.L. 1999. Detecting a salinity plume in a unconfined sandy aquifer and assessing secondary soil salinization using electromagnetic induction techniques, North Dakota, USA. *Hydrogeology Journal*, 7:380-392

Ibáñez, S. 2005. Comparación de la aplicación de distintos modelos matemáticos sobre acuíferos costeros detríticos. Ph. Thesis Universidad de Granada. 304 p.

I.T.G.E. 1991. Investigación hidrogeológica para apoyo a la gestión hidrogeológica en la Cuenca del río Guadalfeo (Granada).

Kafri U., Goldman M. and Lang B. 1997. Detection of subsurface brines, freshwater bodies and the interface configuration in-between by the time domain electromagnetic method in the Dead Sea Rift, Israel. *Enviromental Geology*. 31; 1-2, Pages 42-49.

Ma, Y., Li, Q., Zou, Z., Xia, Z. 2007 Assessment of the hydrological alteration of the Yangtze River IAHS-AISH Publication (311), pp. 546-551

Magán, J.J., Gallardo, M., Thompson, R.B., Lorenzo, P. 2008. Effects of salinity on fruit yield and quality of tomato grown in soil-less culture in greenhouses in Mediterranean climatic conditions Agricultural Water Management

Manchanda, G., Garg, N. 2008. Salinity and its effects on the functional biology of legumes. *Acta Physiologiae Plantarum*, pp. 1-24

W. Martín-Rosales; A. González-Ramón; C. Duque; J.C. Rubio-Campos; M.L. Calvache; M. López-Chicano; J.A. Navarro-García. Repercusión del embalse de Rules en la evolución del nivel piezométrico en el acuífero de Motril-Salobreña. In: Proceedings of Simposio Internacional sobre Tecnología de la Intrusión de agua de mar en Acuíferos Costeros. Almería.

Ministerio de Medio Ambiente, Secretaría de Estado de Obras Hidraúlicas, Dirección General de Obras Hidraúlicas y Calidad de las Aguas, 2005. Presa de Rules en el río Guadalfeo, Vélez de Benaudalla (Granada). Confederación Hidrográfica del Sur de España.

Milliman, J.D., Farnsworth, K.L., Jones, P.D., Xu, K.H. and Smith, L.C. 2008. Climatic and anthropogenic factors affecting river discharge to the global ocean, 1951-2000. *Global and Planetary Change* 62 (3-4), pp. 187-194

Parra, S., Aguilar, F. J., Calatrava, J. 2008 Decision modelling for environmental protection: The contingent valuation method applied to greenhouse waste management. *Biosystems Engineering* 99 (4), pp. 469-477

Pulido-Bosch, A. y Rubio, J.C. 1988. Los acuíferos costeros de Motril-Salobreña. In: Proceedings of Simposio Internacional sobre Tecnología de la Intrusión de agua de mar en Acuíferos Costeros.. *Almuñécar*, pp. 209-238

Shalash, S. 1983 Degradation of the River Nile. *International Water Power and Dam Construction* 35 (7), pp. 37-43

Van Cauwenbergh, N., Pinte, D., Tilmant, A., Frances, I., Pulido-Bosch, A., Vanclooster, M. 2008. Multi-objective, multiple participant decision support for water management in the Andarax catchment, Almeria. *Environmental Geology* 54 (3), pp. 479-489

Wilson, S.R., Ingham M., McConchie, J.A., 2005. The applicability of earth resistivity methods for saline interface definition. *Journal of Hydrology*, 1-12.

Yermiyahu, U., Ben-Gal, A., Keren, R., Reid, R.J. 2008. Combined effect of salinity and excess boron on plant growth and yield Plant and Soil 304 (1-2), pp. 73-87

In: Dams: Impacts, Stability and Design
Editors: Walter P. Hayes and Michael C. Barnes

ISBN 978-1-60692-618-5
© 2009 Nova Science Publishers, Inc.

Chapter 8

ANALYSIS OF FACTORS INFLUENCING THE EXTENT OF HYDROLOGICAL CHANGES OF ANNUAL MAXIMUM AND MINIMUM FLOW DOWNSTREAM FROM DAMS IN QUEBEC

Ali A. Assani[1]*, Martin Matteau[1], Mhamed Mesfioui[2] and Stéphane Campeau[3]

1. Laboratoire d'hydro-climatologie et de Géomorphologie fluviale, Section de Géographie (Hydroclimatology and River Geomorphology Laboratory, Geography Section), Pavillon Léon-Provencher, Université du Québec à Trois-Rivières, 3351 Boulevard Des Forges, Trois-Rivières, Quebec, G9A 5H7, Canada
2. Département de Mathématiques et d'Informatique (Department of Mathematics and Computer Science), Université du Québec à Trois-Rivières, 3351 Boulevard des Forges, Trois-Rivières, Quebec, G9A 5H7, Canada
3. Laboratoire d'analyse des bassins versants, Section de Géographie, Pavillon Léon-Provencher, Université du Québec à Trois-Rivières, 3351 Boulevard Des Forges, Trois-Rivières, Quebec, G9A 5H7, Canada

ABSTRACT

The application of the ecological concept of natural flow regimes to the study of hydrologic impacts of dams necessitated simultaneous analysis of numerous hydrologic variables and identification of the factors influencing the extent of hydrological changes of these variables downstream from dams. Due to some of their weaknesses, the simple and multiple regression methods did not make the most of this concept's contribution to the study of hydrologic impacts of dams. To mitigate these weaknesses, we apply canonical correlation analysis. We correlated 7 hydrologic variables which define the fundamental characteristics of annual maximum and minimum streamflows with 8 explanatory factors for 62 stations for which streamflows are measured downstream from

* Corresponding author: Ali A. Assani, Tel.: (819) 376-5011; Fax: 376-5179; email: Ali.Assani@uqtr.ca.

dams in Quebec. This analysis identified the factors influencing the magnitude of the hydrologic changes in the characteristics of the annual maximum and minimum streamflows downstream from dams in Quebec.

- Streamflow magnitude and frequency are mainly influenced by the watershed size. The changes in annual maximum streamflow are greater for large watersheds (> 10,000 km²) than for small watersheds concerning annual minimum streamflows. This factor may be sufficient to estimate the streamflows downstream from dams deprived of hydrologic data, except for the lowest minimum streamflows.
- The timing of annual maximum streamflows is mainly influenced by the type of regulated hydrologic regime (dam management mode) while the timing of annual minimum streamflows is mainly influenced by the degree of water storage.
- The timing variability of annual minimum streamflows is influenced by the cumulative maximum capacity of the reservoirs in the system (number of dams built on the same watercourse).
- The variability and asymmetry of the magnitude are not influenced by any explanatory factor analyzed.

Keywords: *Canonical correlation analysis, streamflow characteristics, annual maximum and minimum streamflows, dams, Quebec*

I. INTRODUCTION

According to Hart et al. (2002), "Water flow is a master variable that governs the fundamental nature of streams and rivers, so it should come as no surprise that the modifications of flow caused by dams alters the structure and function of river ecosystem". Many studies thus have been dedicated to analysis of the hydrological changes induced by dams (e.g., Batalla et al, 2004; Erskine et al., 1999; Graf, 1999; Higgs and Petts, 1988; Leconte et al., 2001; Magilligan and Nislow, 2005; Page et al., 2005; Peters and Prowse, 2001; Rosenberg et al., 2000; Richter et al., 1998). These studies have the following two main objectives:

- Quantification of the extent of hydrological changes of the hydrologic variables downstream from dams.
- Identification of the factors influencing this extent of the hydrological changes of these hydrologic variables downstream from dams.

The first objective has been studied extensively. The three commonly used methods (monitoring station method, control station method and natural streamflow reconstitution method) to study the impacts of the dams were formulated to achieve this objective. They are based on comparison of the streamflows measured under natural conditions and those released downstream from dams.

However, very few studies exist dedicated to the determination of the factors influencing the extent of hydrological changes downstream from dams (Assani et al., 2005; 2006a; Batalla et al., 2004; Lajoie et al., 2007; Magilligan and Nislow, 2005). These studies used the regression method (simple or multiple) to identify these factors. The hydrologic variables measured downstream from dams thus are correlated with the physiographic and climatic characteristics of the watersheds and the characteristics of the dams. However, these methods present several weaknesses which do not allow them to make the most of the contribution of the natural flow regime concept in analyzing the hydrologic impacts of dams. Indeed, regression is unsuitable when several hydrologic variables (dependent variables) have to be analyzed simultaneously, because it does not account for the effect of intragroup relationships on intergroup relationships. Moreover, due to the relatively high number of hydrologic variables to be analyzed, regression becomes cumbersome to apply. The results are relatively unsynthetic, making their interpretation difficult. In the case of simple regression, not all the explanatory factors are taken into account. Finally, these studies are limited only to the factors influencing the change of magnitude of the streamflows downstream from dams. Other streamflow characteristics are never analyzed. To mitigate all these weaknesses, we propose using canonical correlation analysis for the first time to identify the factors influencing the extent of the hydrological changes of hydrologic variables downstream from dams.

Identification of the factors influencing this extent is interesting for two main reasons.

- It allows determination of the factors that can be taken into account for the development of flood and low flows standards downstream from dams in order to restore and preserve the ecological integrity of the regulated sections.
- It also allows selection of the relevant variables to estimate the streamflows downstream from dams when the streamflow measurements are unavailable.

The purpose of this study is to determine the factors influencing the extent of the hydrological changes of the characteristics annual maximum and minimum streamflows downstream from dams in Quebec and select the most relevant factor to estimate their magnitude. This aspect has never been analyzed in the scientific literature.

II. METHODOLOGY

II.1. Selection of Study Stations and Data Sources

This study pertains to 62 regulated flow stations of the St. Lawrence River (Figure 1). Their watershed sizes range from 211 to 143,000 km² and the maximum capacity of their reservoirs, from 74,000 to 25 billion m³. The selection of these stations was based on the following criteria: continuous streamflow measurement during a period of at least ten years downstream from the dam or at the dam level, structure mainly intended for exploitation of hydroelectric power. The streamflow data analyzed come from Environment Canada's Hydat CD-ROM (2004). The data on the dams were extracted from the *Répertoire des barrages du Centre d'Expertise Hydrique du* Québec (www.cehq.gouv.qc.ca/dams, consulted 03/08/2007). This dam directory contains various information on the identification, administrative category and type of use and on the technical characteristics of each dam (dam height, dam type, dam class, year of dam construction, etc.). These data allowed us to calculate the dam characteristics.

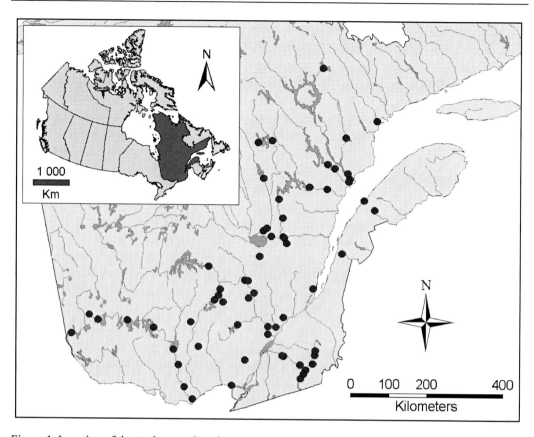

Figure 1. Location of the stations analyzed.

Table 1. Definition of Streamflow Characteristics and Hydrologic Variables

Streamflow characteristics	Hydrologic variables	Code	Calculation mode
Magnitude	Average	M0	Average of the annual maximum and minimum streamflows
Magnitude frequency	Percentile 90	PMF90	Streamflow exceeding 90% of the values
	Percentile 10	PMF10	Streamflow below 10% of the values
Magnitude variability	Coefficient of variation	CVM	Ratio between the standard deviations and their average streamflows
Timing	Average	T0	Average timing (in julian days) of annual maximum and minimum streamflows
Timing variability	Coefficient of variation	CVT	Ratio between the standard deviations and their average timing
Shape of the distribution curve	Coefficient of asymmetry of magnitude	CAS	Pearson's correlation coefficient B

Table 2. Description of Independent Variables

Class of variables	Type of variables	Code	Calculation mode	Units of measure
Physiographic	Watershed size*	DA	-	Km²
	Latitude	Lat	-	(°)
	Longitude	Long	-	(°)
Dam characteristics	Total maximum capacity*	CM	-	m³
	Cumulative maximum capacity*	CCM	∑CM upstream	m³
	Degree of storage	IR	CM/MAF	-
	Degree of cumulative storage	IRC	CCM/MAF	-
Management modes or hydrologic regimes	Inversed flow regime	RR**	-	-
	Homogenization flow regime	RR**	-	-
	Natural-type flow regime	RR**	-	-

MAF = annual average streamflow under pristine conditions. For some stations, MAF was estimated according to the regional equation established by Assani et al. (2007). * = the values were converted into natural logarithms. ** = qualitative variables. These variables were rated from 1 (inversed flow regime) to 3 (natural flow regime).

II.2. Definition of Extreme Streamflow Characteristics and Explanatory Factors

Concerning streamflows, we applied the concept of "natural streamflow regime" (Poff et al., 1997; Richter et al., 1996) to define the six fundamental streamflow characteristics according to the methodology proposed by Assani et al. (2005, 2006a, 2006b). These characteristics and the hydrologic variables that define them are summarized in Table 1.

The explanatory factors have been grouped in three categories (Table 2):

- The factors that describe the physiographic characteristics of the watersheds: watershed size (DA) and the longitude (LONG) and latitude (LAT) of the gauging station located downstream from the dam. We should specify that latitude and longitude also include a site's climatic conditions. We did not have precipitation and temperature data at each station. This is why these data were not analyzed in this study.
- The factors relating to the dam characteristics: the maximum capacity (CM) of a single reservoir, located just upstream from the station, and the sum of the maximum capacities (CCM) of all the other reservoirs located upstream from the station. This last factor is a measurement of the number of reservoirs built on the same watercourse. Apart from these two factors, we have also calculated the IR index, which is the ratio between the maximum capacity of a reservoir (CM) and the natural

mean annual flow (converted in m³) entering the reservoir. This index measures the degree of water storage in a watershed. In case of the presence of numerous reservoirs, an equivalent index (IRC) was calculated, accounting for all the reservoirs located upstream. We should mention that the mean annual flow is little influenced by dams in Quebec (Assani et al., 2007). Its use is therefore justified to calculate the IRC values. In case of absence of data on natural mean annual flow entering a reservoir, these data were estimated by means of the equations established by Assani et al. (2007) for natural rivers in Quebec. This estimate is based exclusively on watershed size, which explains more than 95% of the total variability of mean annual flow.

–   The factors describing the dam management modes (RR). In a series of studies, Assani et al. (2005, 2006a, 2007) identified three management modes. Each of these modes was associated with a specific hydrologic regime (Figure2). Three types of regulated hydrologic regimes were thus described. The first type of regime is the inversion flow regime. It is characterized by maximum streamflows in winter and minimum streamflows in spring. This is the opposite of the natural annual hydrologic cycle of streamflows in Quebec. This regime is generally associated with reservoirs often located upstream from the channels and designed to feed the hydroelectric generating stations located downstream. The second type of hydrologic regime, called the homogenization flow regime, is characterized by streamflows that remain almost constant all year around. Finally, the last type of regime, the natural flow regime, preserves the natural streamflow cycle. However, it is distinguished from the pristine flow regime by a slight increase in streamflows in winter and a slight decrease in streamflows in spring. For statistical analysis purposes, the three types of flow regimes, being qualitative variables, received the following ratings: 1 (the inversed flow regime), 2 (the homogenization flow regime) and 3 (natural-type flow regime). The inversion flow and natural flow regimes include nearly 80% of the stations, divided almost equally, while the homogenization flow regime includes less than 20% of the stations.

II.3. Statistical Data Analysis

To determine the factors influencing the magnitude of the changes in streamflow characteristics downstream from dams, we applied canonical correlation analysis. This is the only method that allows simultaneous correlation of several dependent variables (hydrologic variables) with several independent variables (explanatory factors). The goal of the method is to be able to select the explanatory factors best correlated with the hydrologic variables in order to estimate the hydrologic variables subsequently. Canonical correlation analysis creates pairs of linear combinations between each group of variables (independent variables and dependent variables) called canonical variables, so that the correlation between the variables of the same pair is maximized and so that the correlation between the variables of two different pairs is nil. The analyses were performed with SAS version 8 software and the following description adapted from Vanderpoorten and Palm (1998).

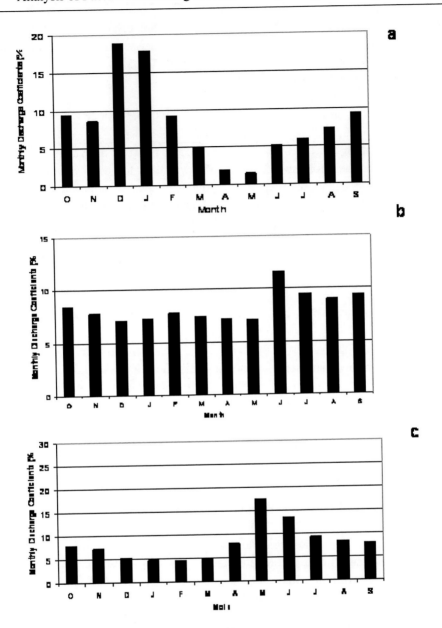

Figure 2. The three regulated regimes observed downstream from dams in Quebec. a = the Inversed flow regime; b = the homogenization flow regime; c = the natural-type flow regime.

Let us consider p dependent variables x_j $(j = 1,...,p)$ and q dependent variables y_k $(k=1, ...,q)$, with x^0_j and y^0_k as their respective standardized variables. For each of these groups of standardized variables, the canonical variables are calculated as follows:

$$w_l = a_{1l}x^0_1 + a_{2l}x^0_{2+...} a_{pl}x^0_p , \qquad (1)$$

$$v_l = b_{1l}y^0_1 + b_{2l}y^0_{2+...} b_{ql}y^0_q \qquad (2)$$

The maximum value of l corresponds to the number of variables of the smallest group.

The canonical correlation coefficients a_{jl} and b_{kl} are calculated according to the two criteria (3) and (4):

$$\text{Corr }(w_l+1, v_l+1) < \text{Corr }(w_l, v_l) \tag{3}$$

$$\text{Corr }(w_l, w_{l'}) = 0, \text{corr}(v_l, v_{l'}) = 0 \text{ and corr}(w_l, v_{l'}) = 0 \text{ if } l' \neq l \tag{4}$$

Linear correlation coefficients are then calculated between the variables of each of the original groups and their corresponding canonical variables to obtain what some authors call structural coefficients, and between the original variables of a group and the canonical variables of the opposite group. Finally, we should mention that some analyzed variables, such as watershed size (BV), were normalized by logarithmic conversion.

III. RESULTS

III.1. Determination of the Factors by Canonical Correlation Analysis

Table 3 presents the values of the canonical correlation coefficients calculated between the extreme streamflow characteristics and the explanatory factors. In the assumption that the sample is not probabilistic, the first three coefficients must be considered statistically significant for annual maximum streamflows while only the first two are statistically significant for annual minimum streamflows. The value of the first correlation coefficient is higher for annual maximum streamflows than for annual minimum streamflows. This means that there is a relatively strong relationship between the streamflow characteristics and the explanatory factors for annual maximum streamflows than for annual minimum streamflows.

Concerning annual maximum streamflows, the structural coefficient analysis reveals that the first canonical variable V1 is strongly correlated (positive correlation) to the average timing of these streamflows (Table 4a). The second canonical variable (V2) is correlated to the three variables that define streamflow magnitude-frequency. Finally, the last canonical variable V3 is not significantly correlated to any hydrologic variable. As for the canonical variables associated with the explanatory factors, the first variable (W1) is correlated to many factors (Table 4b). However, the strongest correlation is observed between W1 and the type of regulated hydrologic regime. The second canonical variable (W2) is strongly correlated (negative correlation) to watershed size (BV). Figure 3a reveals that the variance between the annual maximum streamflows of pristine rivers and regulated rivers seems to be greater for large watersheds (> 10,000 km²) than for small watersheds (< 10,000 km²). Thus, the decrease in streamflow magnitude-frequency downstream from dams is greater for large watersheds than for small and medium-sized watersheds. Finally, the last canonical variable (W3) is moderately correlated to the longitude and cumulative maximum capacity of reservoirs (CC and IRC). It follows that the extent of the hydrological changes of the timing of annual maximum streamflows downstream from dams is strongly influenced by the type of hydrologic regime (management mode). Magnitude-frequency is influenced by watershed size. Finally, the variability of magnitude and timing is weakly influenced by the maximum capacity of the reservoirs upstream from the measurement station.

Table 3. Values of Canonical Correlation Coefficients

	Annual maximum streamflows			Annual minimum streamflows		
	R^2	F	p< F	R^2	F	p< F
1	**0.89085**	**4.29**	**<0.0001**	**0.799434**	**2.72**	**<0.0001**
2	**0.723835**	**2.41**	**<0.0001**	**0.733287**	**1.98**	**0.0018**
3	**0.538512**	**1.62**	**0.0373**	0.521577	1.16	0.2854
4	0.461686	1.32	0.1915	0.411876	0.82	0.6627
5	0.313903	0.88	0.5483	0.251731	0.45	0.9026
6	0.190892	0.58	0.6785	0.120327	0.24	0.9163
7	0.078016	0.33	0.5676	0.071579	0.25	0.6177

The statistically significant coefficients appear in bold.

Table 4a. Structural Coefficients Calculated Between the Annual Maximum Streamflows and the New Canonical Variables

	V1	V2	V3	V4	V5	V6	V7
M0	-0.3997	**0.8175**	-0.2405	-0.1414	-0.0928	-0.2756	-0.0978
PMF90	-0.4102	**0.8283**	0.2700	-0.0765	-0.1467	-0.2122	0.0194
PMF10	-0.1658	**0.6108**	0.1147	-0.5749	0.0273	-0.4626	-0.2026
CVM	0.0470	0.3716	-0.4744	-0.0312	0.0658	0.6983	0.4081
CAS	0.3723	-0.0202	-0.4779	-0.3659	0.6179	0.1032	0.3261
T0	**0.8169**	0.3167	-0.1871	0.3454	0.2708	-0.0663	0.0155
CVT	0.4528	-0.1509	-0.4946	-0.1034	-0.7046	0.1387	0.0338
VE (%)	19.5	28.4	12.5	8.9	14.1	12.2	4.6

VE = Explained Variance.

Table 4b. Structural Coefficients Calculated Between the Explanatory Factors and the New Canonical Variables (Annual Maximum Streamflows)

	W1	W2	W3	W4	W5	W6	W7
RR	**-0.9673**	-0.0162	0.2519	-0.0554	0.0890	0.1614	-0.0437
CM	**0.6475**	-0.0571	0.1438	0.1234	0.4336	-0.1462	-0.5763
CCM	**0.6546**	0.2186	**0.6063**	-0.2898	0.1476	0.2219	-0.0326
DA	-0.0132	**-0.8977**	0.3544	0.2541	0.0412	0.0244	0.0399
Lat	0.4597	0.1993	0.3572	0.7543	0.1663	-0.1078	-0.1147
Long	0.2434	-0.3696	**-0.6067**	-0.4002	0.2072	0.3350	0.3475
IR	**0.6362**	0.3470	-0.0162	0.0221	0.3348	-0.1010	-0.5931
IRC	**0.6550**	0.1830	**0.6213**	-0.2800	0.1495	-0.2232	-0.0311
VE (%)	36.1	14.8	18.4	12.2	5.3	3.5	10.3

VE = Explained Variance

For the annual minimum streamflows, the first canonical variable V1 is strongly correlated to the hydrologic regime (management mode) but weakly correlated to the coefficient of variation of magnitude (Table 5a). The second canonical variable V2 is correlated to magnitude-frequency. Finally, the third canonical variable V3 is correlated to the coefficient of variation of timing of the annual minimum streamflows. The first canonical variable W1, extracted from the group of explanatory factors, is also correlated to several factors (hydrologic regime, maximum capacity, IR and latitude). However, the strongest correlation was observed with the IR index (Table 5b). The second canonical variable W2 is correlated to watershed size and, to a lesser extent, to latitude. It is observed that, contrary to the annual maximum streamflows, the variance of annual minimum streamflows between natural rivers and regulated rivers is greater for small and medium-sized watersheds (< 10,000 km²) than for large watersheds (Figure 3b). Finally, the third canonical variable W3 is correlated to the CCM and IRC indices. In light of this analysis, contrary to the annual maximum streamflows, the timing of annual maximum streamflows is better correlated to the IR index. On the other hand, magnitude-frequency is still correlated to watershed size. Finally, the timing variability of annual minimum streamflows is also correlated to the same factors (CCM and IRC).

Table 5a. Structural Coefficients Calculated Between the Annual Minimum Streamflows and the New Canonical Variables

	V1	V2	V3	V4	V5	V6	V7
M0	0.4263	**0.8771**	0.0450	-0.0438	-0.0133	0.1112	-0.1804
PMF90	0.3094	**0.8176**	0.1541	0.1188	0.0431	0.2235	-0.3823
PMF10	0.4877	**0.6986**	-0.0515	-0.4025	0.2633	0.0557	0.1922
CVM	**-0.5463**	-0.4298	0.3215	0.1331	-0.0456	0.4177	0.4683
CAS	-0.4739	-0.2390	0.3032	0.4305	0.0544	0.1842	0.6357
T0	**0.9278**	0.0781	0.2716	-0.0489	-0.1714	0.1585	-0.0486
CVT	-0.2523	0.1917	**-0.8273**	0.2164	0.1484	0.3826	-0.0016
VE (%)	28	31.6	14	6.1	1.8	6.3	12

VE = Explained Variance.

Table 5b. Structural Coefficients Calculated Between the Explanatory Factors and the New Canonical Variables (Annual Minimum Streamflows)

	W1	**W2**	W3	W4	W5	W6	W7
RR	**0.6353**	0.0275	-0.5400	0.4268	-0.0021	0.2813	-0.2067
CM	**-0.6318**	0.2793	0.2861	-0.1511	-0.2048	0.5229	0.3206
CCM	-0.0577	-0.0360	**0.8855**	0.1581	0.0968	0.3761	0.1885
DA	0.3895	**0.7634**	0.0977	-0.3190	-0.3388	-0.0832	0.1803
Lat	**-0.6089**	**0.5763**	0.3508	0.2306	-0.1019	0.0696	-0.3250
Long	0.1192	-0.0714	-0.0557	-0.4578	0.5257	-0.3037	0.6319
IR	**-0.7813**	-0.0668	0.2507	0.0420	-0.0877	0.4942	0.2619
IRC	-0.0432	-0.0075	**0.8899**	0.1463	0.0842	0.3734	0.1954
VE (%)	24.4	12.6	26.9	7.7	5.8	12.3	10.3

VE = Explained Variance.

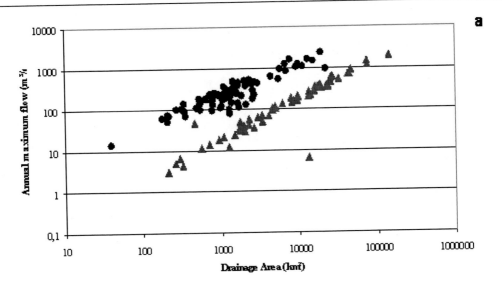

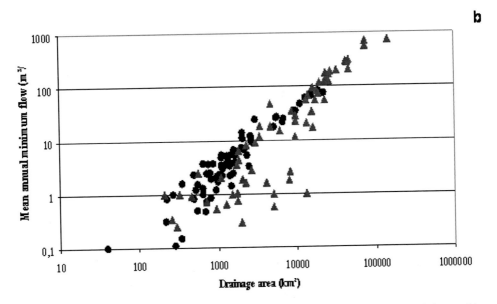

Figure 3. Comparison between the magnitude (average) of annual maximum (a) and minimum (b) streamflows between natural rivers (black points) and regulated rivers (red triangles).

The comparison of the variance values explained by the new canonical variables reveals that, for the group of hydrologic variables, the highest explained variance is associated with V2 both for maximum streamflows and annual minimum streamflows. However, the variance explained by the first two new canonical variables (V1 and V2) is higher for annual minimum streamflows (59.6%) than for annual maximum streamflows (48%). As for the group of explanatory factors, the highest explained variance is associated with V1 for annual maximum streamflows but with V3 for annual minimum streamflows. On the other hand, the total variance explained by the first two new canonical variables becomes lower for annual minimum streamflows (37%) than for annual maximum streamflows (50.9%).

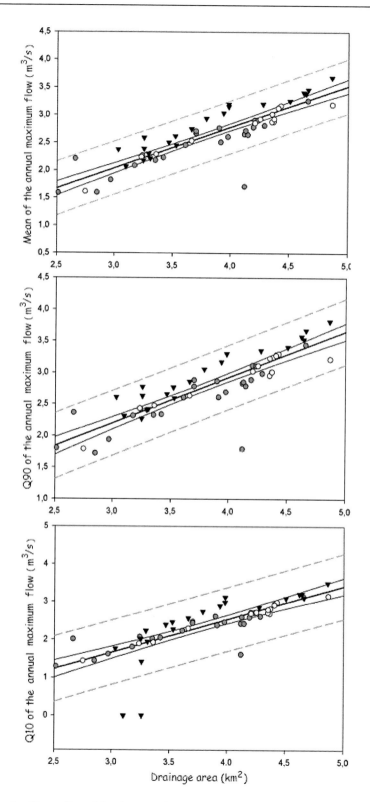

Figure 4. Regression lines adjusted for the magnitude-frequency of annual maximum streamflows. a = Average (M0), b = Percentile 90 (PMF90); c = Percentile 10 (PMF10).

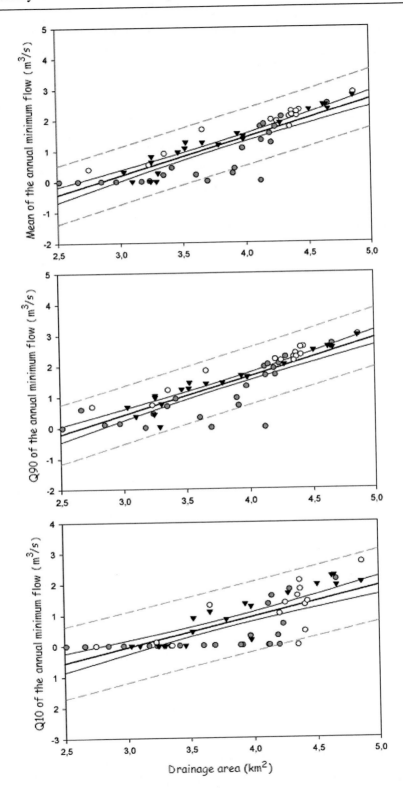

Figure 5. Regression lines adjusted for the magnitude-frequency of annual minimum streamflows. a = Average (M0), b = Percentile 90 (PMF90); c = Percentile 10 (PMF10).

Table 6a. Covariance Analysis Results. Annual Maximum Streamflows

| Factors | Annual maximum streamflows | | | | | |
| | M0 | | PMF90 | | PMF10 | |
	F	P	F	p	F	p
Regimes	0.127	0.881	0.011	0.989	2.014	0.143
Drainage area	347.91	**0.000**	285.32	0.000	219.17	**0.000**
Interaction	0.885	0.418	0.475	0.625	2.73	0.074

F is the Fisher test; p = F test significance threshold.

Table 6b. Covariance Analysis Results. Annual Minimum Streamflows

| Factors | Annual minimum streamflows | | | | | |
| | M0 | | PMF90 | | PMF10 | |
	F	P	F	p	F	p
Regimes	1.03	0.364	0.182	0.834	1.34	0.271
Drainage area	221.62	**0.000**	211.670	0.000	71.63	**0.000**
Interaction	1.86	0.166	0.611	0.546	2.01	0.144

F is the Fisher test; p = F test significance threshold.

III.2. Estimating Annual Maximum and Minimum Streamflows Downstream From Dams

One of the objectives of identification of the factors influencing the extent of hydrological changes of streamflow characteristics downstream from dams is undoubtedly the selection of the relevant factor or factors to estimate the streamflows. At first glance, estimating the streamflows downstream from dams cannot be justified because these streamflows are supposed to be measured daily downstream for all dams during training operations. Consequently, the streamflow data logistically should be available downstream from each dam. However, in the course of our work performed on the environmental impacts of dams in Quebec (Assani et al. 2002; 2005; 2006; 2007), we found that streamflow data is deficient downstream from most dams in Quebec for various reasons. This deficiency thus justifies estimating the streamflows downstream from these dams to the extent that the extreme streamflow estimating methods developed for natural rivers cannot be transposed to regulated rivers (e.g., Anctil and Mathevet 2004; Anctil et al. 1998; 2000; Daviau et al. 2000; Gingras et al. 1994; Grehys 1996a, 1996b; Haché et al. 2002; Javelle et al. 2003; Leclerc and Ouarada 2007; Ouarda et al. 1999a; Ouarda et al. 2001; Pandey and Nguyen 1999; Ribeiro-Corréa et al. 1995). Moreover, as the Nicolet Report (1997) on the Saguenay flood had pointed out, the extreme streamflow data downstream from dams are important to ensure the safety of the human populations living downstream and to prevent the risks related to extreme weather events, in particular. Fortier et al. (2008) showed the absence of correlations between a reservoir's inflow and outflow. This absence of correlation means that the streamflows released downstream from a dam cannot be estimated from the natural streamflows entering a reservoir, because of the influence of factors such as temperature in winter hydroelectric power production.

The results of the canonical correlation analysis revealed that the magnitude-frequency of annual maximum and minimum streamflows downstream from dams is better correlated to watershed size. Consequently, this factor is the only relevant variable to estimate the extreme streamflows downstream from dams deprived of these data. We thus calculated a simple regression between the hydrologic variables defining magnitude-frequency and watershed size. The correlation between streamflows and watershed size is presented in Figures 4 and 5. The values of the coefficients of determination (R^2) range between 0.819 and 0.583. These values are higher for maximum streamflows than for minimum streamflows, except for P90. On the other hand, both for maximum streamflows and for minimum streamflows, the coefficients of determination are lower for P10 than for P90 and the average values. Whatever the case may be, over 90% of the stations are located in the space defined by the prediction intervals. Estimating streamflows thus can be considered satisfactory overall. However, for the P10 values of annual minimum streamflows, the analysis of the graph (Figure5c) reveals that for small watersheds, streamflows do not increase as a function of watershed size. They tend to remain constant. This is explained by the fact that null streamflow values are often observed downstream from small dams in dry years due to major water storage in the reservoirs (Assani et al. 2005). This situation impairs the quality of the estimate of these minimum streamflows. Finally, the disposition of the stations of the three regimes in relation to the regression line and the prediction intervals does not suggest any influence of the management mode on the quality of the estimate even if, at equal watershed size, the streamflows in the inversion flow regime are generally lower than those measured in the natural flow regime. This was confirmed by the covariance analysis, which only showed the influence of watershed size (Tables 6a and 6b).

IV. DISCUSSION AND CONCLUSION

Canonical correlation analysis made it possible to identify the factors influencing the extent of hydrological changes of streamflow characteristics downstream from dams. This analysis shows that the timing of the annual maximum and minimum streamflows, associated with the first canonical variable, is influenced by several factors, principally the type of hydrologic regime (dam management mode) and the reservoir capacity measurement parameters (CM and IR). These changes are mainly associated with the inversion flow regime. This is characterized by winter timing of annual maximum streamflows and spring timing of annual minimum streamflows. Indeed, a large quantity of water is released downstream from the reservoirs in winter for hydroelectric power production. Thus, the floods occur in winter. In spring, water is stored to fill the reservoirs. Consequently, a small quantity of water is released downstream from the reservoirs during this season. The minimum streamflows thus occur in spring. On a pristine river, however, the floods occur in spring when the snow melts and the minimum streamflows occur in winter due to the absence of flow on the watersheds (precipitation falls in the form of snow and does not generate runoff). Moreover, the reservoirs associated with the inversion regime have a greater maximum capacity than other types of reservoirs due to their role intended to supply water to the hydroelectric generating stations located downstream for hydroelectric power production during the cold season (Assani et al., 2005; 2006a).

As for streamflow magnitude-frequency, associated with the second canonical variable, this characteristic is mainly correlated to watershed size. The impact of the dams translates into a decrease in magnitude-frequency both of annual maximum streamflows and of annual minimum streamflows. However, the magnitude of this decrease downstream from dams depends almost exclusively on watershed size. Indeed, for maximum streamflows, the decrease is greater for large watersheds than for small watersheds. The opposite is true for annual minimum streamflows. For large watersheds, a lot of water is released downstream to produce much of the electricity intended to supply Quebec's major cities and for export to the United States and Canadian provinces like Ontario. As for the annual minimum streamflows, the decrease becomes significant for small and medium-sized watersheds because more water is stored in the reservoirs to avoid any water shortage during the winter season. Finally, we showed that watershed size may be sufficient to estimate the maximum and minimum streamflows downstream from dams in Quebec. The use of this variable to estimate streamflows presents the following two major advantages:

- The watershed size values are easy to obtain.
- The estimate is based on a simple regression equation. However, the quality of streamflow estimates by this method is not as good for the lowest annual minimum streamflows (P10). We thus consider the use of non-linear regression to improve the quality of these streamflow estimates.

The timing variability of annual minimum streamflows is mainly correlated to the CCM and IRC indices, two factors that measure the number of reservoirs built on the same watercourse. The presence of many reservoirs tends to mitigate the timing variability of annual maximum and minimum streamflows. A stronger regulating effect thus occurs.

In their study devoted to multiple regression analysis of the factors influencing the extent of hydrological changes of hydrologic variables downstream from dams in the United States, Magilligan and Nislow (2005) observed that the magnitude of annual minimum streamflows was influenced by watershed size, while the magnitude of annual maximum streamflows was influenced by latitude. The timing of annual maximum streamflows was influenced by the annual average temperature. Thus, a difference from Quebec is observed, especially regarding the influence of climate. This factor seems to have less impact in Quebec, due to the fact that climate spatial variability is relatively low there compared to the United States. Consequently, its influence on the hydrologic variables becomes negligible.

REFERENCES

Anctil, F.; Coulibaly, P. *J. Climate*. 2004, 17, 163-173.
Anctil, F.; Mathevet, T. *Can. Water Res. J.* 2004, 29, 47-58.
Anctil, F.; Martel, N.; Hoang, V.D. *Can. J. Civ. Eng.* 1998, 25, 360-369.
Anctil, F.; Larouche, W.; Hoang, V.D. *Water Qual. Res. J. Canada*. 2000, 35, 125-146.

Assani A.A.; Buffin-Bélanger T.; Roy, A.G. *Rev. Sci. Eau.* 2002, 15, 557-574.

Assani A.A.; Gravel, E.; Buffin-Bélanger, T.; Roy, A.G. *Rev. Sci. Eau.* 2005, 18, 103-127.

Assani, A.A.; Stichelbout, E.; Roy, A.G.; Petit, F. *Hydrol. Process.*, 2006a; 20, 3485-3501.

Assani, A.A.; Tardif, S.; Lajoie, F. *J. Hydrol.* 2006b, 328, 753-763.

Assani, A.A.; Lajoie, F.; Laliberté, C. 2007. *Rev. Sci. Eau.* 2007, 20, 127-146.

Batalla, J.; Gomez, C.M.; Kondolf, G.M. *J. Hydrol.* 2004, 290, 117-136.

Black, A.R.; Rowan, J.S.; Duck, R.W.; Bragg, O.M.; Clelland, B.E. *Aquat. Conserv.: Mar. Freshwat. Ecosyst.* 2005, 15, 427-446.

Daviau, J-L.; Adamowski, K.; Patry, G.G. *Hydrol. Process.* 2000, 14, 2731-53.

Environnement Canada 2004. National Water Data Archive (Hydat), CDROM v. 2.04, January 2004.

Erskine W.D.; Terrazzolo N.; Warner, R.F. *Rivers: Res. and Mgmnt.* 1999, 15, 3-24.

Fortier C.; Assani A.A.; Mesfioui, M.; Roy, A.G. *J. Hydrol.* 2008 (submitted)

Gingras, D.; Adamowski, K. 1994. *Can. J. Civ. Eng.* 994, 21, 856-862.

Graf, W.L. *Water Resourc. Res.* 1999, 35, 1305-1311.

GREHYS. *J. Hydrol.* 1996a, 186, 63-84

GREHYS. *J. Hydrol.* 1996b 186, 85-103.

Haché, M.; Ouarda, T.B.M.J.; Bruneau, P.; Bobée, B. 2002. *Can. J. Civ. Eng.* 2002, 29, 899-910.

Hart D.D. ; Johnson T.E. ; Bushaw-Newton K.L. ; Horwitz R.J. ; Bednarek A.T. ; Charles D.F. ; Kreeger D.A. ; Velinsky. *Bioscience.* 2002, 52, 669-681.

Higgs, G.; Petts, G. 1988. *Regul. Rivers : Res. and Mgmnt.* 1988, 2, 349-368.

Javelle, P.; Ouarda, T.B.M.J.; Bobée, B. *Hydrol. Process.* 2003, 17, 3717-3736.

Lajoie, F.; Assani, A.A.; Roy, A.G.; Mesfioui, M. *J. Hydrol.* 2007, 334, 423-439.

Leclerc, M.; Ouarda, T. B.M.J. *J. Hydrol.*, 2007, 343, 254-265.

Leconte, R.; Pietroniro, A.; Peters, D.L.; Prowse, T.D. *Regul. Rivers: Res. and Mgmnt.* 2001, 17, 51-65.

Magilligan, F.J.; Nislow, K.H. 2001. Long-term changes in regional hydrologic regime following impoundment in a humid-climate watershed. *JAWRA.* 2001, 37, 1551-1569.

Magilligan, F.J.; Nislow, K.H. *Geomorph.* 2005, 71, 61-78.

Nicolet, R. 1997. Rapport de la Commission scientifique et technique sur la gestion des barrages. 350p + annexes.

Page K., Read A., Frazier P., Mount N. *River Res. Applic.* 2005, 21, 567-578.

Peters D.L.; Prowse T.D. 2001. *Hydrol. Process.* 2001, 15, 3181-3194.

Ouarda, T.B.M.J.; Rasmussen, P.F.; Cantin, J.-F.; Bobée, B.; Laurence, R.; Hoang, V.D.; Barabé, G. *Rev. Sci. Eau.* 1999, 12, 425-448.

Ouarda, T.B.M.J.; Girard, C.; Cavadias, G.S.; Bobée, B. 2001. *J. Hydrol.* 2001, 254, 157-173.

Pandey, G.R.; Nguyen, V.T.V. *J. Hydrol.* 1999, 92-101.

Poff, N.L.; Allan, J.D.; Bain, M.B.; Karr, J.R.; Prestegaard, K.L.; Richter, B.D.; Sparks, R.E.; Stromberg, J.C. *Bioscience.* 1997, 47, 769-784.

Ribeiro-Corréa, J.; Cavadias, G.S.; Clément, B.; Rousselle, J. *J. Hydrol.* 1995, 175, 71-89.

Richter, B.D.; Baumgartner, J.V.; Powell, J.; Braun, D.P. *Conserv. Biol.* 1996, 10, 1163-1174.

Richter, B.D.; Baumgartner, J.V.; Braun, D.P.; Powell, J. *Regul. Rivers: Res. and Mgmnt.* 1998, 14, 329-340.

Rosenberg, D.M.; McCully, P.; Pringle, C.M. 2000. *Bioscience.* 2000, 50, 746-751.

Suen, J.P.; Eheart, J.W. *Water Resourc. Res*. 2006, 42, W03417.
Vanderpoorten, A.; Palm, R.. *Hydrobiologia*. 1998, 386, 85-93.

In: Dams: Impacts, Stability and Design
Editors: Walter P. Hayes and Michael C. Barnes

ISBN 978-1-60692-618-5
© 2009 Nova Science Publishers, Inc.

Chapter 9

GROWING MORE RICE WITH LESS WATER IN ASIA: IDENTIFYING AND EXPLORING OPPORTUNITIES THROUGH SYSTEM OF RICE INTENSIFICATION

Abha Mishra and V. M. Salokhe[*]

Agricultural Systems and Engineering
Asian Institute of Technology
Bangkok, Thailand

ABSTRACT

Due to ever-increasing population, water scarcity, and pressing issues of environmental sustainability, Asian rice farmers are under considerable pressure for sustainable increase in rice production by using less water. It is widely believed that an increase in the water use efficiency through integrated crop management holds promise of increased yields and water productivity. The so-called System of Rice Intensification (SRI) is attracting favorable attention of farmers and governments in Asia and elsewhere. It is assumed that a healthier and larger root system can be induced by using "SRI principles" in a water-limiting environment giving positive impacts on grain yield. The cultural practices that characterize SRI includes rapid and shallow transplanting of younger seedlings, at wider spacing and maintaining alternate wet and dry condition or preferably just moist conditions during the vegetative stage. This chapter reviews the biological mechanisms of water-saving agriculture and its relation to SRI cultural practices. It presents some research findings on the rice plant's adaptive trait which could be utilized to manage crops under limited water application. In addition, the details of on-farm studies carried out in some of the rice growing countries of Southeast Asia using a participatory action research approach are included in this chapter, which asserts the need for integrating science, people and policy makers for better and sustainable water management in Asia and elsewhere.

---

[*] E-mail: salokhe@ait.ac.th

Keywords: agricultural water management, System of Rice Intensification (SRI), root growth, Participatory Action Research (PAR), Farmers' Field School (FFS)

INTRODUCTION

Asian farmers and rice growers in particular, are under considerable pressure to increase rice production sustainably by using less water. Major drivers are the growing population, water scarcity, reduced investment in irrigation infrastructure and pressing issues of environmental sustainability. While the water shortage in this region is severe, water use efficiency of rice is very low and is further lowered at the small farms which are supposed to contribute 75% of the additional food needed over the next decade. It is widely believed that an increase in the water use efficiency through integrated crop management holds promise to increase yields, improve water productivity leading to narrow the yield-gap experienced in these farmers' field in most of the developing countries of Asia. One, in particular, the so-called System of Rice Intensification (SRI) is attracting favourable attention of farmers and governments in Asia and elsewhere to mitigate theses challenges.

SRI, which is mostly empirical method of growing rice, is defined as a set of basic management practices (Stoop et al., 2002; http://ciifad.cornell.edu/sri) that includes:

- Transplanting very young seedlings (10-15 days old)
- Rapid and shallow transplanting of one or two seedlings with wider spacing (30 x 30 cm or 40 x 40 cm)
- Practicing alternate wetting and drying (AWD) during vegetative phases (or keeping the soil moist but not continuously saturated)
- Applying organic manure as much as possible

The SRI concepts, developed by Fr. Laulaniè in the 1980's in Madagascar, was intended to enable resource-limited farmers to obtain higher yields with less water and without primarily relying on external inputs for yield improvement. It sought to achieve this by altering the ways that rice plants, soil, water and nutrients are managed (Stoop et al., 2002). The cultural practices that characterize SRI such as rapid and shallow transplanting of younger seedlings, wider spacing and reduced irrigation water usage are all amenable to farmer experimentation and adaptation to suit local conditions. Thus, SRI encourages farmer participation in devising practical ways of growing a healthy crop in a sustainable manner.

In SRI, there are synergistic effects among the recommended practices, giving more increase in yield in water limiting environment when all are used together rather than used singly and separately. The synergistic effect of components of SRI practice rekindles some latent issues that are still not given adequate attention. One such critical issue is the contribution of root vigor towards yield (Mishra et al., 2006) especially in water limiting environment. It is proposed that a healthier, larger and prolonged active root system can be induced by using the SRI method in a water-limiting environment (Satyanarayana et al., 2007; Mishra et al., 2006), which is assumed to boost grain yield per plant. However, there is limited formal research activity investigating SRI and thus many have challenged the yield gains reported (McDonald et al., 2006). Given that agronomic practices are radically different from those promoted during the Green Revolution, the SRI provides unique opportunities for farmers, researchers, and extension workers to engage and be actively involved in exploring these newer agronomic practices. In particular, it is interesting to explore how these new agronomic practices can change the rice plant's phenotype and yield performance and how this can be achieved with reduced irrigation inputs so that farmers can obtain 'more crop per drop'. Certainly, this requires on-station experiments and on-farm studies to clarify various bio-technical issues, as well as to develop location specific technology for better adaptability of SRI practices.

This chapter reviews some of the biological mechanism of water-saving agriculture in relation to SRI cultural practices. The results of on-station research on some of the management practices of SRI, which have got major attention in Asia, is also discussed. In addition, the chapeter presents the outcome of on-farm studiescarried out using participatory action research in Cambodia and Thailand with aim to find synergies among various agronomic management practices in a given socio-economic conditions and to build capacity of the participants to manage a healthy rice crop in a water limiting environment.

BIOLOGICAL MECHANISM OF WATER-SAVING AGRICULTURE AND SRI CULTURAL PRACTICES

A water-saving agricultural system refers to integrated farming practices that are able to sufficiently use rainfall and irrigation facilities for improved water use efficiency (Shan, 2002). The scientific measures in a water-saving agricultural system include spatial and temporal adjustment of water resources, effective use of natural rainfall, rational use of irrigation water and increased plant *water use efficiency* (WUE). In agricultural practices, several factors need to be take into account, namely, (i) the quantity, quality, spatial and temporal distribution of water resources, (ii) the establishment of cultivation practices aimed at reducing water consumption as a result of reshaping the existing farming structure and cropping systems in line with the current distribution pattern of water resources, (iii) sufficient manpower and equipment for the research, development, production, supply and maintenance of water-saving materials, spare parts, instruments and facilitates, (iv) relevant laws and statutes concerning water management to be enacted, formulated and perfected, and (v) a special campaign to enhance the public's water-saving awareness (Deng et al., 2003).

Among all, the establishment of newer cultivation practices, aimed at reducing water consumption as a result of reshaping the existing farming practices with better on-farm management can make significant contribution to 'grow more rice with less water' in Asia. A good understanding of factors limiting and/or regulating yield may provide researchers opportunity to identify and select physiological and breeding traits that increase plant WUE and drought tolerance under water-scarce conditions, where as better understanding and knowledge of soil-plant water relationship can help farmers to manage their rice crop with optimal water use. The details here basically deal with the physiological regulation of biological water-saving which might be possible to achieve using SRI cultural practices. The details of efforts to facilitate farmers for better understanding of soil-plant-water relationship through participatory action research conducted in Cambodia and Thailand for better on-farm water management are discussed too.

Regulating Photosynthesis by Increased Root Activity

Under field conditions there is a parabolic relationship between photosynthesis and transpiration (Wang and Liu, 2003), such that transpiration increases as photosynthesis increases but continues to increase when photosynthesis reaches a maximum so that ultimately transpiration efficiency decreases. At these levels, transpiration can be controlled without affecting the rate of photosynthesis if root activity is kept high (root activity is á-naphthylamine-oxidizing activity of roots which is correlated with higher chlorophyll content of the leaves, slower leaf senescence, erect leaves and single leaf photosynthetic ability). Jiang et al. (1985) suggested that if root activity is kept high, photosynthesis doesn't decline much in the afternoon. By this way transpiration efficiency might be increased to enhance WUE. Based on that, a 'broken irrigation' was introduced in the past to control tillering and to maximize efficiency at the harvest by regulating photosynthesis and by changing the sink-source relationship. In this method, after flooding the field, water intake was cut so as to let the water level go down gradually until the soil surface appeared. The field surface was exposed to air for three to five days, and then flooded again. A similar water management practice was recommended under SRI which suggests that these manipulations might have influence on root activity to regulate photosynthesis (Mishra et al., 2006). An experiment was conducted on rice in semi-field condition at Asian Institute of Technology (AIT), Thailand in 2006 on varying water regimes. Root activity and chlorophyll content of upper and lower leaf were studied under three water regimes:

i) Intermittent flooding (IFI) – Pots were maintained with IFI water regimes. The water depth was maintained at 5 cm everyday for 12 days, then drained for three days and again reflooded to the same depth of ponded water. Water was drained from the bottom of the pot by drainage hole and collected in container and was stored in a refrigerator during pot drying and returned to the pots when reflooding was done. Three, three-days-drying period were provided at 19, 34 and 50 DAT followed by flooded water treatment (5 cm water depth continuously) until maturity.

ii) Intermittent flooding (IFII) – In other pots, similar procedure like IFI was followed five times at 19, 34, 50, 66 and 82 DAT followed by flooded water until maturity.

iii) Continuous flooded (CF) – 5 cm depth of ponded water was maintained until maturity.

The experiment revealed that there was positive correlation between chlorophyll content of lower leaves and root activity in all water regimes (Figure 1). Significant and positive correlation was also found between chlorophyll content of flag leaves and duration of grain fillings (Figure 2). It was observed that senescence of lower leaf and flag leaf was delayed under intermittent irrigation applied during vegetative stage compared to continuous flooded condition and continuous intermittent irrigation (Figure 3). This delay was associated with higher root activity, and higher biomass production along with higher grain weight (Table 1). Therefore, it seems that intermittent irrigation recommended in SRI might decrease irrigation water requirement and would contribute towards higher yield when soil nutrients are not the limiting factor.

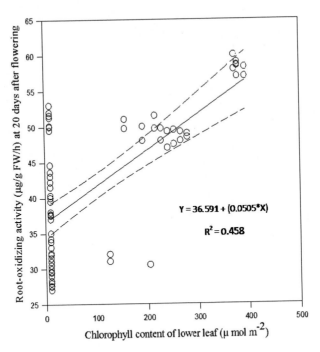

Figure 1. Linear regression slope between chlorophyll content of lower leaf (x) and root oxidizing activity at 20 days after flowering (y). Dotted lines show the 95% confidence level.

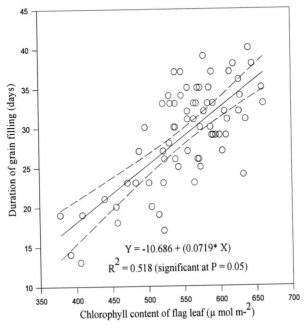

Figure 2. Linear regression slope between chlorophyll content of flag leaf and duration of grain filling at 95% confidence level. Dotted lines show the 95% confidence level.

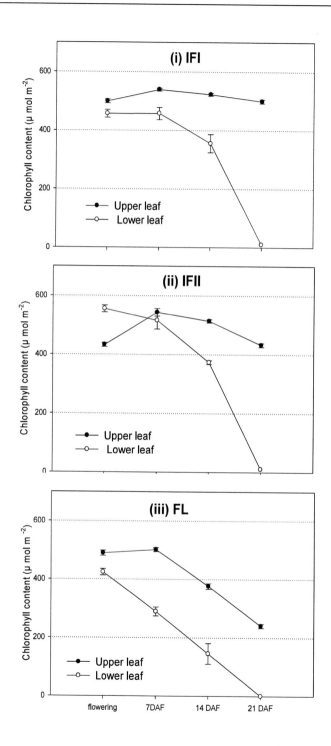

Figure 3. Changes in chlorophyll content of upper (flag leaf) and lower (3rd leaf) leaf of rice plant during reproductive phase maintained under three water regimes: IFI (intermittent drying for three times), IFII (intermittent drying for five times) and NF (nonflooded). Error bars show *s.e.* (n=2). For each replicate, three leaves were used for measurement. DAF denotes days after flowering.

Table 1. Changes in root activity, plant biomass, and grain weight under varying waterregimes in pot study conducted at AIT. (IFI – intermittent drying for three times; IFII – intermittent drying for five times; and CF – continuous flooding)

Treatment	Root oxidizing activity (μg/g FW/h) at flowering	Root oxidizing activity (μg/g FW/h at 20 days after flowering	Total biomass (g/plant)	Dry grain weight per pot (g)
IFI	63.40 ± 0.74	42.28 ± 0.57	341.71 ± 10.22	165.74 ± 4.07
IFII	53.10 ± 0.35	33.10 ± 0.33	215.73 ± 7.67	116.46 ± 2.19
CF	62.34 ± 0.56	30.12 ± 0.22	248.14 ± 9.75	101.75 ± 7.28

Means and *s.e.* at DF = 4, P< 0.05.

Regulating Plant-Water Relationship by Modifying Root Morphology

A deep root system is synchronous with more water uptake from the soil and better performance under drought. A deep and healthy root system is not only correlated with better water uptake but also influence yield physiology by regulating cytokinin production.

Cytokinin is regarded as the most important senescence-retarding plant hormone (Faiss et al., 1997). It is greatly regulated by environmental conditions of rhizosphere such as nitrogen availability, soil moisture condition, root mass, root length density etc. Among all, root length density and root mass are important variables for characterizing temporal trends in the water relations of rice especially when the water supply is scarce. The amount of water available to plant depends on the relative root length density and the ability of roots to absorb water from the soil.

Experiments conducted by Matsuki and Katsutani (1940) have shown larger root weight in lower soil layers under limiting moisture conditions. Baba (1977) also found that non-flooded soil condition in the early and mid stages of rice growth promoted the appearance of xeromorphic traits with increased root mass, and was consistent with physiological adaptation to drought conditions without compromising the yield. This was possible due to larger root growth both at upper and lower soil depths, and osmotic adjustment at early growth stage that helped plant to maintain photosynthesis with increased WUE. The simplest way to increase rooting depth and root distribution of crops is to increase the duration of the vegetative period. This may be achieved by sowing earlier or delaying flowering. SRI practice recommends transplanting of younger (12-15 days old) seedlings with wider spacing. This helps plants to have prolonged vegetative periods along with better canopy growth and enhanced canopy photosynthesis by avoiding shading effects. Intermittent irrigation or preferably 'just moist' soil condition at vegetative stage further helps plants to grow more roots with higher root length density at deeper soil layer. Yang et al. (2004) showed that intermittent irrigation increased the root length density, active absorption area, root oxidation ability, and nutrient uptake in rice.

Taking it to further, an experiment was conducted in 2006 at AIT in nursery seedbed and after transplanting to understand the roots and shoot characteristics of younger and older seedlings grown in wet and dry seedbed and their contribution to the production of tillers and dry matter. It was found that the root length density was affected by the age of seedlings, seedbed management and by water regimes at early growth stage. Flooded soil favored root length density at shallow soil depth mainly for older seedlings (Table 2). This could be due to preference of shoot growth over root growth in older seedlings and dominance of NH_4^+ in the soil solution which under reduced environment remains mostly in upper soil layer (Sah and Mikkelsen, 1983). In contrast, non-flooded soil improved root growth in the subsoil layer, but more in younger seedlings compared to older ones. The better uptake of N by younger seedlings grown in a dry seedbed was also seen. It was possible due to higher root length density and greater number of lateral roots that helped better acquisition of nutrients from the soil. Younger seedlings raised in a dry seedbed appeared to be the most suitable management practice to achieve higher tiller production along with higher biomass production even under non-flooded soil condition due to better root growth (Figure 4 and 5). This adaptive root trait could be exploited to manage rice crops under limited water application without compromising the grain yield (Mishra and Salokhe, 2008).

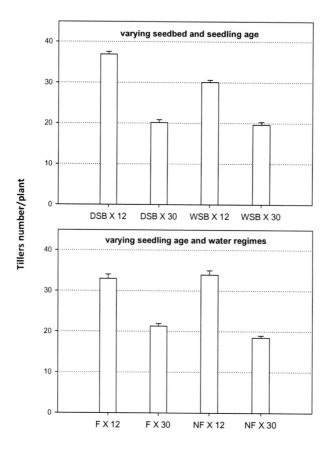

Figure 4. Plant's tiller development at 45 days after transplanting with interaction effect of different seedbed (DSB = dry seedbed and WSB = wet seedbed), seedling age (12 = 12 days old and 30 = 30 days old) and water regimes (F= flooded and NF = nonflooded). (N = 16). Error bars show *s.e.*

Table 2. Effects of seedbed management, seedling age, and water regimes on root characteristics and N uptake by rice plant at 45 days after transplanting in pot study conducted at AIT

Seedbed	Water regimes	RLD (cm cm^{-3}) upper soil layer		RLD (cm cm^{-3}) sub-soil layer		Total RLD (cm cm^{-3})		N content in plant shoot (mg pot^{-1})	
		12 days	30 days	12 days	30 days	12 days	30 days	12 days	30 days
DSB	F	5.78±0.14	4.55±0.08	1.97±0.13	1.30±0.04	7.75	5.84	321.75±4.83	253.88±4.82
DSB	NF	5.35±0.18	4.25±0.16	2.55±0.14	1.20±0.08	7.90	5.46	312.75±7.68	181.88±2.45
WSB	F	5.21±0.13	4.26±0.17	1.77±0.12	1.66±0.11	6.98	5.92	274.50±9.04	205.50±2.53
WSB	NF	5.14±0.12	3.22±0.20	1.92±0.12	1.79±0.07	7.06	5.01	243.75±5.00	184.63±3.09

RLD = Root length density. Mn and N content in shoot were expressed on dry weight basis.

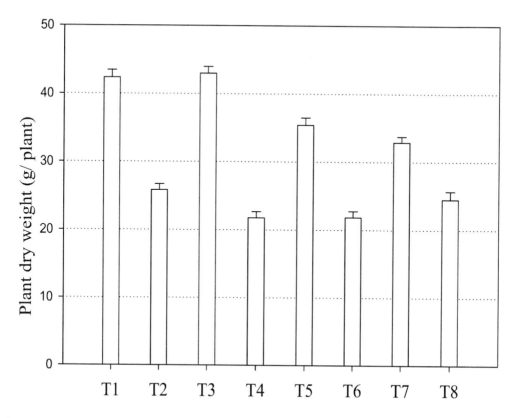

Figure 5. Plant biomass at 45 days after transplanting with effect of different seedbed, seedling age and water regimes. X axis shows the different treatments (T1= DSB+12+F; T2 = DSB+30+F; T3 =DSB+12+NF; T4 =DSB+30+NF; T5 = WSB+12+F; T6 = WSB+30+F; T7 = WSB+12+NF; T8 = WSB+30+NF). (N = 8) Error bars show *s.e.* (See figure 4 for notations used).

Based on the above findings it may be stated that a healthier, larger and prolonged active root system can be induced by using the SRI cultural practices in a water-limiting environment. However, these remain to be tested over different locations with different variety before making any general conclusions, due to limited formal research activity investigating SRI systematically.

ENGAGING FARMERS, RESEARCHERS AND EXTENSION WORKERS THROUGH SRI FOR BETTER ON-FARM MANAGEMENT FOR *'MORE CROP PER DROP'*

Above reviews and some initial research findings indeed provide some opportunity to manage rice crop by optimizing soil-plant-water relationship in a water limiting environment. However, given the agro-ecological and socio-economic diversity of rice production system, there is a need to develop knowledge-intensive and location-specific technology to realize the merit of SRI crop management principles.

In recent year, there has been an increase in the number of publications about SRI However, there is no disagreement that there still exists a significant gap between on-research-station yields and those obtained by most Asian farmers in their own paddy fields (Stoop and Kassam, 2005; Mc.Donald et al., 2006). This gap and current economic exigencies along with water scarcity provide incentives for farmers to explore the potential of novel options offered by SRI. It offers a set of management practices that farmers can evaluate, adapt and then adopt to meet their own local requirements. Moreover, it relies minimally on external inputs and maximally on farmers acquiring and using new knowledge and skill.

Yield increase at farmers' field under SRI management compared to conventional practice has been reported repeatedly (Koma, 2002, Anthofer, 2004). Even though more and more farmers are actively engaged in evaluating and adopting SRI practice, many knowledge gaps remain, and robust investigations might help to understand the SRI phenomena. In particular, such process of inquiry should focus on 'location-specific' management practices for optimizing soil-plant-water relationship for better yields with minimal water application. Given that agronomic practices are radically different from those promoted during the Green Revolution; the SRI provides unique opportunities for farmers, researchers, and extension workers to engage-in and be actively involved- in exploring these newer agronomic practices in general, and growing healthier root systems, in particular. Such collaborative investigation efforts require common and viable platforms at the grass-root level. Farmers' Field Schools (FFS), an approach developed and promoted by FAO to educate rice farmers on Integrated Pest Management (IPM) in Asia (Kenmore, 1991) offer suitable platforms for such investigative partnerships to explore SRI.

Keeping these concepts and facts in mind, a collaborative Participatory Action Research (PAR) initiative was set up in Prey Veng, Cambodia during 2005/6 under overall coordination of the National IPM Program of Ministry of Agriculture, Forestry and Fisheries with funding support from Regional IPM Program of Food and Agriculture Organization of the United Nations.

Two major factors – water and planting were identified by participants. Subsequently, a field trial was designed at Prey Veng province of Cambodia. The main factor was water regimes -- flooded (FL), alternate wet and dry (AWD), and just moist (JM), and the sub-factor was planting densities – six seedlings/hill with narrow spacing (6; 15 x15 cm), single seedling with narrow spacing (1; 15 x 15 cm) and single seedling with wider spacing (1; 30 x 30 cm). In 'FL' treatment, 5 cm water depth from the soil surface was maintained until panicle initiation stage and then 10 cm water depth was followed until 7 days before harvesting. In 'JM' moisture regimes, soils were kept continuously moist by irrigating plots with 1-1.5 cm water depth from the soil surface and re-applied when the top-soil surface started to dry up. In 'AWD' moisture regimes, intermittent irrigation was followed with initial 5 cm water depth from the soil surface and re-irrigated when the soil had developed fine cracks. In both 'JM' and 'AWD' water treatments; irrigation was stopped 20 and 25 days respectively before harvesting. –The single seedling with wider spacing under alternate wet and dry water regimes was treated as SRI where as six-seven seedlings/ hill with 15 x 15 cm spacing under flooded water regimes was treated as conventional practice.

After transplanting, observations on above-ground plant parts were made during regular Agro-Ecosystem Analysis (AESA) sessions that included yield attributing characters (per hill and per m^2) such as plant height, number of tillers/hill, number of leaves/tiller, leaf color at vegetative and reproductive stage, and panicle length, number of productive tiller/m^2, number of grains/panicle, number of filled grains/panicle, thousand grain weight, and yield/plot at harvest time. In addition, observations were made on the water level in each plot, insect pest populations, and on weather conditions. To correlate above-ground plant parameters and their related findings to the below-ground plant parts – roots were observed at seedling, tillering, flowering and harvesting stage by developing root AESA.

Based on the agronomic and AESA results, the participants involved in this action research concluded that:

- Using less water under 'just moist' field condition more number of productive tillers could be produced (Figure 6).
- Less water (under JM and AWD) at later growth stage accelerate grain filling (Figure 7), so this gives further opportunity to produce higher yields with less water inputs.
- A single seedling/hill is better than 6-7 seedlings/hill in terms of getting higher grain yield with better economic returns (Fig 8). Generally, farmers use 35-45 kg seed/ha for transplanting whereas for using the single seedling method – 6-7 kg seed/ha would be sufficient.
- At vegetative stage, the rice plant grows equally better in just moist or alternate wet and dry water regimes compared to those grown under continuous flooded regimes; hence, water inputs could possibly be reduced.

Results clearly demonstrated that SRI effect on rice yield is not governed by single factors but there are many factors which act simultaneously and in a synergistic mode. Such system understanding can only be possible when farmers are able to understand these interaction effects. IPM farmers, who have gone through season-long Farmers Field Schools (FFS), have developed such ecosystem thinking and management skills through discovery-learning (van de Fliert and Braun, 2002). In this trial, the IPM principles were followed by using "Agro-ecosystem Analysis" (AESA) -- a useful training tool, that focuses on farmers making regular field observations and collecting data on the plant and its micro-environment. Data are then analyzed and summarized on a so-called Agro Ecosystem Analysis (AESA) poster followed by plenary group discussions with the aim to arrive at informed crop management decisions. In this SRI investigation, AESA helped farmers to discover the fundamental causal relationship between roots and shoots of the rice plant which varied under different planting densities and water regimes. In particular, these investigations assisted farmers to understand why ensuring the growth of healthy root systems has the potential of boosting yields in water limiting environment. This had truly been innovative and resulted into the development of adapted FFS training curricula and learning exercises on growing healthy root systems.

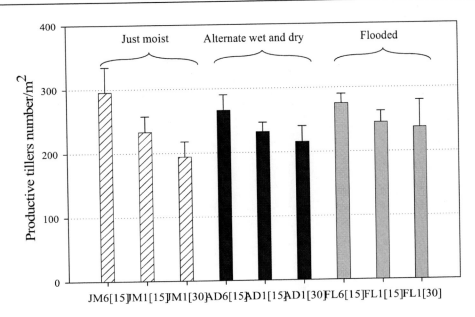

Figure 6. Highest number of productive tillers under JM water regimes in participatory action research (PAR) trial conducted in Cambodia. JM6[15] represents six seedlings with 15 x15 cm spacing under just moist (JM) water regimes; JM1[15] is single seedling with 15 x 15 cm spacing under JM; JM1[30] is single seedling with 30 x 30 cm spacing under JM water regimes. Similarly AD and FL are alternate wet and dry and flooded water regimes respectively. Error bars show *s.e.*

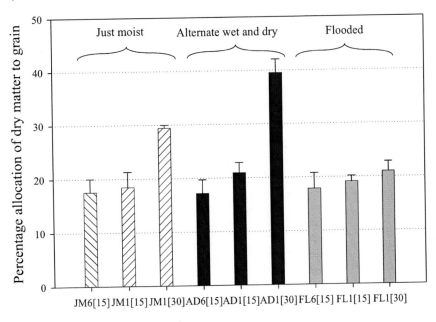

Figure 7. Higher percentage of dry matter allocation to the grain in single seedling with 30 x 30 cm spacing grown in nonflooded water regimes (JM and AWD) in PAR trial conducted in Cambodia. (See Figure 6 for details of symbol on X axis). Error bars show *s.e.*

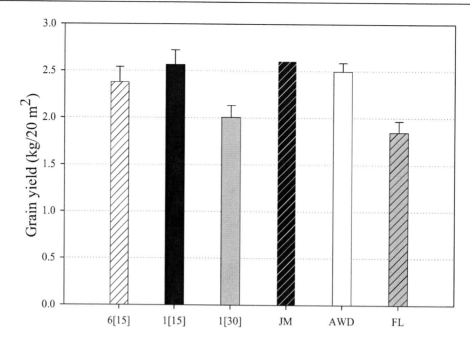

Figure 8. Grain yield at 14% moisture content in different water regimes and planting densities in PAR trial conducted in Cambodia. (See Fig. 6 for details of symbol on X axis). Error bars show *s.e.* Please note that FL (flooding) is simulated for conventional practice whereas, nonflooded (AWD or JM) is simulated for SRI method. There was no interaction between water regimes and planting densities for grain yield.

A similar collaborative enquiry into the water productivity issues of the transplanted rice under the ambience of SRI were carried out with a group of farmers, Non Government Organization (NGO) and Government Organization (GO) personnel in Ban Chaeng, District, at Samart, Roi-Et, Thailand using a participatory action research with a funding support from the Consultative Group of International Agricultural Research (CGIAR) through 'Challenge Program for Water and Food' grant with project Title "Increasing Water use efficiency in Rice using Green mulch under SRI management practice". 'Water use in rice' was discussed with the participating farmers and non-formal education trainees during weekly FFS conducted for 18 weeks. Two experiments were carried out during first season in Wet season 2006, and in experiment-1, where the two water regimes i.e., Just moist (JM) was compared with the farmers' practice (flooding), no significant difference in crop yields were noticed. The JM produced similar rice yield per unit area with less supplementary irrigation. In experiment 2, where SRI practice was followed in combination with green mulch, SRI and Mung Bean combination was proving to be the best among all other tested bean intercropping, thereby providing high foliage and ground cover as green mulch to the rice crop grown under SRI method. The results of experiments are shown in Figures 9-12. The results clearly showed that higher water use efficiency and rice yield obtained under SRI + green mulch (e.g. Mung bean) under just moist water regimes, and performed better than any existing farmer's practice. Also the results confirmed the benefit of younger seedling transplant (14 days) over older seedling (30 days) in another set of trial.

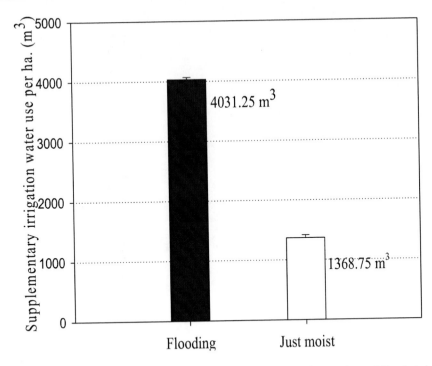

Figure 9. Total volume of supplementary irrigation water used under just moist and flooded rice cultivation system in PAR trial conducted in Thailand. Please note that this information is compiled to show the difference of water-use in two systems of rice; flooding (traditional system) and Just moist i.e., non flooded water regimes (SRI method). Error bars show *s.e.*

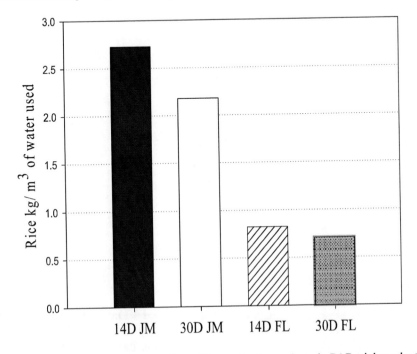

Figure 10. Water productivity for two tested conditions of water regimes in PAR trial conducted in Thailand. Please note that the 30 days old seedling and flooded water regimes (30D FL) are simulated as conventional farming practice used in the experiment for comparison purposes.

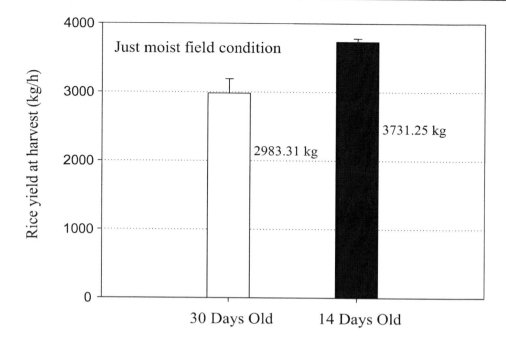

Figure 11. Rice yield under just moist (JM) condition in PAR trial, Thailand. 14 days old seedling performed better over 30 days old seedling under similar water and other management conditions. Error bar shows *s.e.*

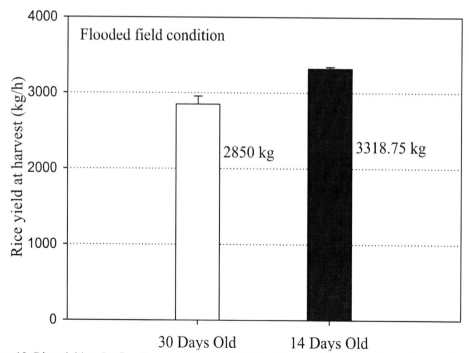

Figure 12. Rice yield under flooding condition in PAR trial, Thailand. 14 days old seedling performed better over 30 days old seedling under similar water and other management conditions. Error bars show *s.e.*

Yields obtained in either of the location in Cambodia and Thailand exceeded the average rice productivity in Prey Veng and Roi-Et province respectively. Indeed, it confirmed that SRI focuses on good crop husbandry and could be seen as a sound alternative management practice to minimize the yield gap experienced in most of the farmers's field in Asia by optimizing inputs use such as water and seed.

The major difference between SRI and the current 'input-driven' rice farming is that, SRI encourages manipulation of plant's aerial and soil environment by changing planting density, water and soil management rather relying on external physical inputs, thus encourages optimization of interaction effect of genetic endowment (G) and environment (E) often termed as 'G x E' effect, and results into healthy plants and healthy roots leading to higher yield. Such optimization requires skills for manipulating internal resources. This could be possible only when farmers are able to understand and appreciate the importance of soil-plant-water relationship for better production.

Hence, given the nature of SRI's focus on system thinking and farmer management skills, empowered IPM farmers appear ideally suited for exploring SRI and capturing potential benefits of such explorations. Given the widening gap between potential farm yield and actual farm yield in Asia (McDonald et al., 2006), such investigations and strengthening of farmers' management and inquiry skills are more urgently needed than ever.

CONCLUSION AND POLICY IMPLICATIONS

The need to grow more rice with less water in a more sustainable production environment is a big future challenge for Asian farmers and supporting R&D systems. SRI appears to provide a good opportunity to explore the contribution of "hidden-half" to yield through basic research by addressing soil-plant-water relationship. Even so, the practical constraints in managing favorable interaction of plant-soil-water environment should not be underestimated. However, these constraints could be minimized if SRI is used as a heuristic vehicle for linking rice research with participatory farmer empowerment programs. The SRI should be seen as a unique opportunity to integrate science with the society for sustainable development to address the 'rice demand with less water' in an integrated manner by focusing on improving farmers' production management skills and involving them in on-farm participatory research.

The lessons learned from the work reported in this chapter include:

- Based on the findings from on- station research trial, it can be assumed that it is possible to increase rice yield by enhancing the physiological efficiency of a rice plant by increasing root activity under limited water supply. Further research should be conducted to unlock the contribution of other roots traits which are strongly linked to the biomass production and grain yield.
- The better SRI results obtained by Cambodian or Thailand farmers suggest that SRI agronomic practices do make sense and deserve further on-farm investigation and promotion within agricultural training/extension programs that focus on developing farmers' production management skills.

- Informed management is needed to realize the synergistic effect of soil-plant-water relationship for better on-farm SRI adaptation. The FFS approach of farmer education appears most suitable for developing such farmer management skills.
- Finally, there is a need for further sensitization of researchers and scientists towards the potential benefits to be captured from more participatory and action oriented research approaches. SRI appears to be ideally suited for such research initiatives.

ACKNOWLEDGEMENTS

Research trials conducted at AIT were partly supported by a grant from the Asia Rice Foundation, USA, 2005. Funding and logistic support from FAO regional vegetable IPM Program for participatory action research on SRI and Soil Ecology in Cambodia is greatly acknowledged. We sincerely thank CGIAR for providing funds for participatory action research set up in Thailand.

REFERENCES

Anthofer, J. (2004). The Potential of the System of Rice Intensification (SRI) for Poverty Reduction in Cambodia. In *"International Agricultural Research for Development"*. *Berlin*, October 5-7.

Baba, I. (1977). Effects of water stress on the physiology and the growth of paddy rice plants in relation to the generation of ethylene. *Japanese Journal of Crop Science.* 46 (1):171-172.

Deng, X. P., Shan, L., Zhang, S.Q., Kang, S. Z. (2003). Outlook on plant biological water-saving strategies. In: *Proceeding of KRIBB Conference on Environmental Biotechnology,* October 21–23. Daejeon, Korea.

Faiss, M., Zalubilova, J., Strnad, M., and Schmulling, T. (1997). Conditional transgenic expression of the IPT gene indicates a function for cytokinins in paracrine signaling in whole tobacco plants. *Plant Journal.* 12: 401- 415.

Jiang, C. Z., Hirasawa, T., and Ishihara, K. (1985). Eco-physiological characteristics of two rice cultivars, photosynthetic rate, water conductive resistance and interrelationship between above and underground parts. *Japanese Journal of Crop Science.* 57: 132-138.

Kenmore, P. E. (1991). 'How rice farmers clean up the environment, conserve biodiversity, raise more food, make higher profits in Indonesia's integrated pest management - A model for Asia' FAO Inter-country Program for the Development and Application of Integrated Pest Control in Rice in South and South-east Asia. (FAO: Metro Manila Philippines).

Koma, Y. S. (2002). Experiences with the system of rice intensification in Cambodia. In: Uphoff, N., Fernandes, C.M., Longping, Y., Jiming, P., Rafaralahy, S. and Rabenandrasana, J. (Ed.): *Assessment of the system of rice intensification* (SRI). Proceedings of an International Conference, Sanya, China, April 1-4, 2002, CIIFAD, Cornell, USA, 83-85.

Matsuki, G. and Katsutani, S. (1940). Some chemical investigations on the drought damage. 1. Soil moisture and plant growth. *Journal of Science and Soil Manure*, Japan. 14: 279-288.

McDonald, A. J., Hobbs, P. R., and Riha, S. J. (2006). Does the system of rice intensification outperform the conventional best management practices? A synopsis of the empirical record. *Field Crop Research.* 96 (1): 31-36.

Mishra, A and Salokhe, V. M. (2008). Seedling characteristics and the early growth of transplanted rice under different water regimes. *Experimental Agriculture* 44 (3): 1-19.

Mishra, A; Whitten, M; Ketelaar, J. W. and Salokhe,V. M. (2006). 'The system of rice intensification (SRI): a challenge for science, and an opportunity for farmer empowerment towards sustainable agriculture'. *International Journal of Agricultural Sustainability.* 4(3):193-212.

Sah, R. N. and Mikkelsen, D. S. (1983). Availability and utilization of fertilizer nitrogen by rice under alternate flooding. II: Effects on growth and nitrogen use efficiency. *Plant and Soil.*, 75: 227-234.

Satyanarayana, A., Thiyagarajan, T. N., Uphoff, N. (2007). Opportunities for water saving with higher yield from the system of rice intensification. *Irrigation Science.* 25: 99-115.

Shan, L. (2002). Development of tendency on dry land farming technologies. *Agricultural Science of China.* 1: 934–944.

Stoop, W. A., and Kassam, A. H. (2005) The SRI controversy: a response. *Field Crops Research.* 91: 357-360.

Stoop, W. A., Uphoff, N., and Kassam, A. H. (2002). A review of agricultural research issues raised by the system of rice intensification (SRI) from Madagascar: Opportunities for improving farming systems for resource-poor farmers. *Agricultural Systems.* 71: 249-274.

Van de Fliert, E. and Braun, A. R. 2002. Conceptualizing integrative, farmer participatory research for sustainable agriculture: From opportunities to impact. *Agriculture and Human Values* 19: 25–38.

Wang, H. X., Liu, C. M. (2003). Experimental study on crop photosynthesis, transpiration and high efficient water use. *Chinese Journal of Applied Ecology*, 14: 1632–1636.

Yang, C., Yang, L., Yang, Y., and Ouyang, Z. (2004). Rice root growth and nutrient uptake as influenced by organic manure in continuously and alternately flooded paddy soils. *Agricultural Water Management,* 70: 67-81.

In: Dams: Impacts, Stability and Design
Editors: Walter P. Hayes and Michael C. Barnes

ISBN 978-1-60692-618-5
© 2009 Nova Science Publishers, Inc.

Chapter 10

A NOVEL MEMORY BASED STRESS-STRAIN MODEL FOR RESERVOIR CHARACTERIZATION

M. Enamul Hossain, S. Hossein Mousavizadegan,
Chefi Ketata and M. Rafiqul Islam

Department of Civil and Resource Engineering,
Dalhousie University
Sexton Campus, 1360 Barrington Street,
Halifax, NS B3J 2X4, Canada

ABSTRACT

The fluid memory is the most important yet most neglected feature in considering fluid flow models, since it represents the history of the fluid and how it will behave in the future. This paper introduces a stress-strain model where all the probable properties have been incorporated with viscous stresses. The derived mathematical model introduces the effect of temperature, the surface tension, pressure variations and the influence of fluid memory on the stress-strain relationship. The part of the stress-strain formulation related to the memory is taken into account and we obtain the variation of it with time and distance for different values of α. The notation α is shown the effect of memory and varied $0 \leq \alpha < 1$. The zero value shows no memory effect while the unity values of α shows the most extreme case of the effect of memory. The fluid memory effects as a function of space and time are obtained for the fluid in a sample oil reservoir. The dependency of fluid memory is considered to identify its influence on time. As pressure is also a function of space, the memory effects are shown in space with pressure gradient change. The computation indicates that the effect of memory cause a nonlinear and chaotic behavior for stress-strain relation. This model can be used in reservoir simulation and rheological study, well test analysis, and surfactant and foam selection for enhanced oil recovery.

NOMENCLATURE

A_{yz} = Cross sectional area of rock perpendicular to the flow of flowing fluid, m^2

c = total compressibility of the system, $1/pa$

E = activation energy for viscous flow, KJ/mol

h_f = height in temperature gradient (height between the two points along the y-direction) or in other words, thin film thickness of flowing fluid, m

k = initial reservoir permeability, m^2

K = operational parameter

L = distance between production well and outer boundary along x direction, m

M_a = Marangoni number

p = pressure of the system, N/m^2

p_i = initial pressure of the system, N/m^2

q_x = fluid mass flow rate per unit area in x-direction, $kg/m^2 - s$

q_i = Au = initial volume production rate, m^3/s

R = universal gas constant, $kJ/mole - k$

T = temperature, °K

t = time, s

ξ = a dummy variable for time i.e. real part in the plane of the integral

u = filtration velocity in x direction, m/s

u_x = fluid velocity in porous media in the direction of x axis, m/s

y = distance from the boundary plan, m

ϕ = porosity of fluid media, m^3/m^3

σ = surface tension, N/m

α = fractional order of differentiation, dimensionless

α_D = thermal diffusivity, m^2/s

μ = fluid dynamic viscosity, $Pa-s$

μ_0 = fluid dynamic viscosity at reference temperature T_0, $Pa-s$

μ_T = fluid dynamic viscosity at a temperature T, $Pa-s$

τ = shear stress, Pa

τ_T = shear stress at temperature T, Pa

$\dfrac{du_x}{dy}$ = velocity gradient along y-direction, $1/s$

ΔT = $T_T - T_0 = T_D - T_A$ = temperature difference (Figure 1), °K

ρ_o = density of the fluid at reference temperature T_0, kg/m^3

$p(x, t)$ = fluid pressure, pa

η = ratio of the pseudopermeability of the medium with memory to fluid viscosity, $m^3 s^{1+\alpha}/kg$

$\left|\dfrac{\partial \sigma}{\partial T}\right|$ = the derivative of surface tension σ with temperature and can be positive or negative depending on the substance, $N/m\text{-}K$

INTRODUCTION

The impact of viscosity in reservoir is directly related to oil production. The mobility of the crude oil increases by lowering the fluid viscosity and consequently, it enhances the oil recovery. A fluid can not resist the shear forces and starts to move and deform regardless of the amount of the shear forces. Its shape will change continuously as long as the force is applied. The relationship between the rate of strain and the shear stress specifies the type of fluid from viscosity points of view. Typically, viscous stresses within a fluid tend to stabilize and organize the flow, whereas excessive fluid inertia tends to disrupt organized flow leading to chaotic turbulent behavior.

The Newton law of viscosity is based on a linear relationship between the viscous stress and the rate of strain. A fluid that satisfies this law is called the Newtonian fluid. The coefficient of proportionality is known as the viscosity that depends only on temperature and pressure and also the chemical composition of the fluid if the fluid is not a pure substance. For a Newtonian fluid, the viscosity does not depend on the forces acting upon it at all shear strain rate. Water, some light-hydrocarbon oils, air and other gases are Newtonian fluids. In contrast, a fluid that has not a well-defined viscosity is called the non-Newtonian fluid. The viscosity of such a fluid changes with the applied forces and strain rates. The Newton model considers the parallel plate concept in fluid flow. The roles of surface tension or interfacial tension on the viscosity of a fluid, and memory of a fluid have been ignored in this model.

In the reservoir, the structure of the formation media is important to consider because the dependency of pore size and microstructure influences the oil flow. Pressure and temperature are the influential factors in flow criteria in porous media where pressure is the most dominant parameter (Hossain et al. 2008). The strain rate in porous media is affected by the velocity of the formation fluid that depends on the pressure variation and in some respect on temperature. This may leads to a non-Newtonian behavior of the crude oil. However, the deviation of the crude oil behavior from a Newtonian fluid is dependent on the formation temperature and also the composition of the crude oil. Crude oil is the outcome of living things which existed millions of years ago like fossils in nature. The American Petroleum Institute (API) defines crude oil as "a substance, generally liquid, occurring naturally in the earth and composed mainly of mixtures of chemical compounds of carbon and hydrogen with or without other nonmetallic elements such as sulfur, oxygen, and nitrogen". It is a complicated mixture of hydrocarbons, with a varying composition depending on its source and pathway. It contains different functional groups, such as paraffins, aromatics, napthenes, resins, and asphaltenes, which may cause dramatic change in the rheological behavior due to the variation of temperature and pressure.

Surface tension is one of the most important fluid properties that govern the fluid flow. Marangoni effect for the fluid layer is considered to represent the effect of surface tension. The Marangoni effect is a phenomenon of interfacial turbulence provoked by surface tension gradients that might be induced by gradients in temperature, concentration and surface charge through the interface (D'Aubeterre et al., 2005). These researchers have given an extensive review of Marangoni effect in heat and mass transfer process. This effect efficiently enhances the rate of mass transfer across the interface. The importance of the Marangoni convection on heat and mass transfer processes in porous media is known by the researchers because of the development of the instabilities at the interface during these two processes. So, the understanding of the behavior of Marangoni effect is very important from both theoretical and practical point of view. This effect is also important in many separation processes such as distillation, absorption and extraction because the convective flows due to the Marangoni effect and other phenomena can lead to increase mass transfer and interfacial turbulence.

Lyford et al. (1998a) concluded that the use of surfactant such as aliphatic alcohol increases the oil recovery in porous media which is due to Marangoni effect. This effect makes oscillations in the droplets trapped in the pores. The cause of oil recovery is the Marangoni interfacial turbulence induced by concentration gradients of solute established by the mass transfer between the trapped droplets and the continuous phase inside the porous media. As a result, Lyford et al. (1998b) proposed that these effects must be characterized by the Marangoni number (M_a), which relates the surface tension gradient with concentration; instead of by the Capillary number (Ca). Moreover, Bragard and Velarde (1998) argued that it is important to understand the instabilities of droplets and bubbles induced by the Marangoni effect. He established that the Marangoni effect is the physicochemical motor which transforms energy in flow and is based on the presence of temperature and concentration gradients, the fluid viscosity and diffusivity.

The stress-strain relation should be time dependent and consider the effect of pathway and fluid memory. In nature, groundwater, oil, gas and other fluids behave according to the pathway traveled by every particle. The conventional approach in reservoir engineering has focused on the permeability of solid and semi-solid structures encountered by the flow of the fluid. What it doesn't do is follow these pathways. Fluid memory is an approach to factor this back in by switching the frame of reference from the external observation of flow to that of matter molecules within the flow. The literature to date has yet to conceptualize fluid memory in a comprehensive way. The particularity and uniqueness of memory is that its definition varies with different combinations of any given fluid and its particular medium which has posed the greatest difficulty. Memory itself is a function of all possible properties of the given fluid and its medium over time. Fluid materials have some common properties but some have special properties that are remarkable in the behavior when some forces are applied on them. In some nonlinear, incompressible and viscous fluids, they possess some peculiar characteristics that lead to think about something else on their properties especially for viscous fluids. The phenomenon that describes these special characteristics is remarked as memory in fluid. There are very limited studies in the literature that describe this phenomenon clearly. However, an extensive review is done here which leads to introduce the notion of memory of fluids.

Jossi et al. (1962) developed a correlation to calculate the viscosity of the pure fluid in terms of the pressure, temperature and properties of the chemical species. Their correlation is suitable to reduced density in the range 0.1 and 3.0. This correlation is valid for the viscosity of nonpolar substances. Because of the inadequacy of current theories in accounting for memory, some authors also developed non-local flow theories (e.g. Hu and Cushman, 1994), using general principles of statistical physics under appropriate limiting conditions from which the classical Darcy's law is derived for saturated flow. Starov and Zhdanov (2001) studied the effective properties of porous media. They correlated viscosity and a resistance coefficient (1/permeability) using Brinkman's equations. They also established the relation between the porosity and viscosity. Abel et al. (2002) studied the boundary layer flow and heat transfer of a visco-elastic fluid immersed in a porous medium over a non-isothermal stretching sheet. They concluded that the fluid viscosity is a function of temperature. Their case study involved to get the effect of fluid viscosity, permeability parameter and visco-elastic parameter for various situations. Their finding is that the effect of fluid viscosity parameter decreases the wall temperature profile significantly when flow is through a porous medium. Further, the effect of permeability parameter decreases the skin friction on the sheet. Brenner (2005) reviewed the Newton's law of viscosity. He found the role of the deviatoric stress tensor in the Navier–Stokes equation for the case of compressible fluids, both gaseous and liquid. He proposed the replacement of velocity gradient term, fluid's mass-based velocity in Newton's law of viscosity by the fluid's volume velocity (volume flux density). He has shown how stress tensors are related with density of a fluid.

Ciarletta et al. (1989) concerned with variational questions about the linearized evolution equations for an incompressible fluid whose viscosity exhibits a fading memory of the past motions. Nibbi (1994) had determined the expressions of some free energies related to viscous fluids with memory. He also considered the quasi-static problem connected with a viscous fluid with memory. Broszeit (1997) dealt with the numerical simulation of circulating steady isothermal flow for liquids with memory. Caputo (1999) investigated in some geothermal areas where the fluids may precipitate minerals in the pores of the medium, leading to diminished sizes. He modified the Darcy's law by introducing a memory formalism represented by a derivative of fractional order simulating the effect of a decrease of the permeability with time. He pointed out some influential parameters that may govern the pore size. The change of pore size influences permeability changes. He also pointed out that permeability diminishes with time and the effect of fluid pressure at the boundary of the pore network on the fluid flow through the medium is delayed and that the flow occurs as if the medium has a memory. Eringen (1999) developed nonlocal theory of memory dependent micropolar fluids with orientational effects. Orientational and nonlocal effects near the walls change viscosity drastically if polymeric fluids squeezed in microscopic sizes. Li et al. (2001) investigated air bubble in non-Newtonian fluids. They identified two aspects for the first time as central to interactions and coalescence: (i) the stress creation by the passage of bubbles, and (ii) their relaxation due to the fluid's memory. Their results show complex nonlinear dynamics, from periodic phenomena to deterministic chaos. Recently, Hossain and Islam (2006) have presented an extensive review of literatures on this memory issue. They critically reviewed almost all the existing models and showed theirs limitations. They also showed how memory of a fluid and media play a great role on stress-strain relation.

The effects of all parameters are considered to obtain a comprehensive model for the stress-strain rate relation for crude oil behavior in the reservoir formation. The effects of surface tension, temperature, pressure, and the fluid memory are taken into account. The effect of surface tension is considered through the application of Marangoni number (M_a)

that explains the role of surface tension. The effect of memory is explained with a dominant variable, the fractional order of differentiation (α), and a constant, ratio of the pseudo-permeability of the medium with memory to fluid viscosity (η). The value of α is changed to identify the effects of memory of the fluid. The results are presented in a graphical form to demonstrate the effect of fluid memory. The results show that the effect of memory causes a nonlinear relationship between memory part of the model and time and space. This nonlinearity is due to the dependency of pressure to the fluid velocity. However, there also exists a chaotic and unpredictable effect on the variation of the fractional order of differentiation.

THEORETICAL DEVELOPMENT

If a tangential force F is acting on the upper surface of a fluid element, ABCDEFGH, in a porous media, AIJKLMNO, in the x-direction, the fluid element will deform as shown in Figure 1. The deformation is caused by shearing forces which act tangentially to a surface, CDEF.

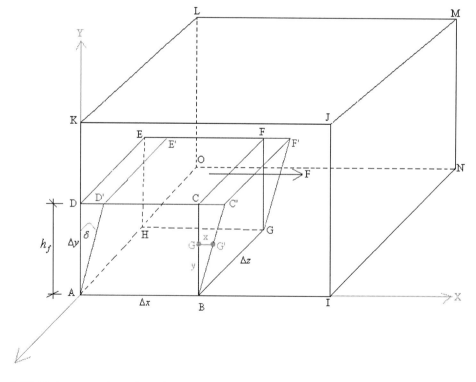

Figure 1. Fluid element under a shear force.

The shear stress and the rate of deformation are normally expressed by the Newton's law of viscosity which can be written mathematically as;

$$\tau = \mu \frac{du_x}{dy}$$

(1)

Now for at any temperature T and in the x-direction, Newton's law of viscosity can be written down into the form;

$$\tau_T = \mu_T \frac{du_x}{dy}$$

(2)

However, this relation is not valid in many cases such as crude oil behavior in porous media.

Effect of Temperature on Fluids

There is some molecular interchange between adjacent layers in liquids but as the molecules are so much closer than in gasses the cohesive forces hold the molecules in place much more rigidly. This cohesion plays an important role in the viscosity of liquids. Increasing the temperature of a fluid reduces the cohesive forces and increases the molecular interchange. Reducing cohesive forces reduces shear stress, while increasing molecular interchange increases shear stress. Because of this complex interrelation the effect of temperature on viscosity has been described by various models depending on industrial applications (Recondo et al., 2005). In this study, the Arrhenius model is used to describe the viscosity dependence on temperature (Avramov, 2005; Recondo et al., 2005; Zuritz et al., 2005; Haminiuk et al., 2006; Gan et al., 2006)

$$\mu_T = \mu_0 e^{\left(\frac{E}{RT}\right)}$$

(3)

Putting the value of μ_T of equation (3) into equation (2) gives;

$$\tau_T = \left[\mu_0 \, e^{\left(\frac{E}{RT}\right)} \right] \frac{du_x}{dy}$$

(4)

High pressure can also change the viscosity of a liquid. As pressure increases the relative movement of molecules requires more energy. Hence viscosity increases, which influences the stress-strain relationship. This effect has been ignored by former researchers.

Effect of Surface Tension and Pressure on Fluids

However, in order to obtain more representative values of the Marangoni effect, it is necessary to isolate the effect from other phenomena such as the natural convection produce by gravity and the presence of surfactants or other agents that can disturb the concentration gradients. Therefore, in the present study, the initial temperature has been considered as 298°K at the plan of ABGH and a certain temperature, T °K at the plan of CDEF along the y-direction (see Figure 1). If the changes in surface tension are produced by temperature gradients, the Marangoni number is expressed as follows (Das and Bhattacharjee, 1999; Haferl and Poulikakos, 2003; Kozak et al., 2004; D'Aubeterre et al., 2005):

$$M_a = \left|\frac{\partial \sigma}{\partial T}\right| \frac{h_f \Delta T}{\alpha_D \mu_0}$$

(5)

Putting the value of μ_0 into Equation (4) and gives:

$$\tau_T = \left[\left(\frac{\partial \sigma}{\partial T} \frac{h_f \Delta T}{\alpha_D M_a}\right) \times e^{\left(\frac{E}{RT}\right)}\right] \frac{du_x}{dy}$$

(6)

If we consider molecularly thin film hydrodynamic fluid behavior in rock matrix, the influence of the fluid discontinuity across the fluid film thickness on the total fluid mass flow through the contact is determined by an operational parameter, K, which is defined as (Zhang and Lu, 2005):

$$K = \frac{\partial p/\partial x \times h_f^2}{6\mu_0(1-\zeta)(u_a - u_b)}$$

If we consider no slip condition and there is no lower contact surface velocity, the above equation can be written as:

$$K = \frac{\partial p / \partial x \times h_f^2}{6 \mu_0 u_x}$$

$$h_f^2 = \frac{6 K \mu_0 u_x}{\partial p / \partial x}$$

It should be mentioned here that in the present molecularly thin film hydrodynamic flow, when the value of the operational parameter K is high, i.e., over 0.1, the influence of the fluid discontinuity across the fluid film thickness on the total fluid mass flow through the contact is significant. When the value of the operational parameter K is low, i.e., close to zero, this influence is negligible (Zhang and Lu, 2005). Putting the value of film thickness, h_f into Equation (6) becomes:

$$\tau_T = \left(\frac{\partial \sigma}{\partial T} \frac{\Delta T}{\alpha_D M_a} \right) \times \left(\frac{6 K \mu_0 u_x}{\partial p / \partial x} \right)^{1/2} \times e^{\left(\frac{E}{RT} \right)} \frac{d u_x}{dy} \tag{7}$$

Effect of Fluid Memory on Viscosity

To incorporate the fluid memory criterion with stress relation, the mass flow rate in porous media can be represented by the following equation (Caputo, 1999). If the direction of flow is in x-direction, the equation can be written as:

$$q_x = -\eta \rho_0 \left[\frac{\partial^\alpha}{\partial t^\alpha} \left(\partial p / \partial x \right) \right] \tag{8}$$

where

$$\partial^\alpha \{ p(x,t) \} / \partial t^\alpha = [1 / \Gamma(1 - \alpha)] \int_0^t (t - \xi)^{-\alpha} [\partial p(x, \xi) / \partial \xi] d\xi \text{ with } 0 \le \alpha < 1$$

Equation (8) can be written as:

$$u_x = -\eta \left[\frac{\partial^\alpha}{\partial t^\alpha} \left(\partial p / \partial x \right) \right]$$

where, u_x is the fluid velocity in porous media. Putting the value of u_x onto Equation (7) yields

$$\tau_T = \left(\frac{\partial \sigma}{\partial T}\frac{\Delta T}{\alpha_D M_a}\right) \times \left(-\frac{6K\mu_0\,\eta\left[\dfrac{\partial^\alpha}{\partial t^\alpha}(\partial p/\partial x)\right]}{\partial p/\partial x}\right)^{1/2} \times e^{\left(\frac{E}{RT}\right)}\frac{du_x}{dy}$$

$$(9)$$

where

$$\partial^\alpha\{\partial p(x,t)/\partial x\}/\partial t^\alpha = [1/\Gamma(1-\alpha)]\int_0^t (t-\xi)^{-\alpha}\left[\partial\left(\frac{\partial p}{\partial x}\right)\middle/\partial\xi\right]d\xi$$

or,

$$\partial^\alpha\{\partial p(x,t)/\partial x\}/\partial t^\alpha = [1/\Gamma(1-\alpha)]\int_0^t (t-\xi)^{-\alpha}\left[\frac{\partial^2 p}{\partial\xi\partial x}\right]d\xi$$

As a result, Equation (9) becomes

$$\tau_T = \left(-\frac{6K\mu_0\eta[1/\Gamma(1-\alpha)]\int_0^t(t-\xi)^{-\alpha}\left[\dfrac{\partial^2 p}{\partial\xi\partial x}\right]d\xi}{\partial p/\partial x}\right)^{1/2} \times \left(\frac{\partial\sigma}{\partial T}\frac{\Delta T}{\alpha_D M_a}\right) \times e^{\left(\frac{E}{RT}\right)}\frac{du_x}{dy}$$

$$\tau_T = \left(\frac{\partial\sigma}{\partial T}\frac{\Delta T}{\alpha_D M_a}\right) \times \left(-\frac{6K\mu_0\eta\int_0^t(t-\xi)^{-\alpha}\left[\dfrac{\partial^2 p}{\partial\xi\partial x}\right]d\xi}{\Gamma(1-\alpha)\dfrac{\partial p}{\partial x}}\right)^{0.5} \times e^{\left(\frac{E}{RT}\right)}\frac{du_x}{dy}$$

$$(10)$$

$$\tau_T = (-1)^{0.5} \times \left(\frac{\partial\sigma}{\partial T}\frac{\Delta T}{\alpha_D M_a}\right) \times \left\{\frac{\int_0^t(t-\xi)^{-\alpha}\left[\dfrac{\partial^2 p}{\partial\xi\partial x}\right]d\xi}{\Gamma(1-\alpha)}\right\}^{0.5} \times \left(\frac{6K\mu_0\eta}{\dfrac{\partial p}{\partial x}}\right)^{0.5} \times e^{\left(\frac{E}{RT}\right)}\frac{du_x}{dy}$$

The above mathematical model provides the effects of the fluid and formation properties in one dimensional fluid flow and this model may be extended to a more general case of 3-Dimensional flow for a heterogeneous and anisotropic formation. However, this may be concluded that, although other causes such as heterogeneity, anisotropy, and inelasticity of the matrix may be invoked to interpret certain phenomena, the memory mechanism could help in interpreting part of the phenomenology. It should be mentioned here that the first part of the Equation (10) is the effects of surface tension, second part is the effects of fluid memory with time and the pressure gradient, the third part is the effects of pressure, viscosity, pseudo-permeability, the fourth part is the effects of temperature on stress-strain equation and the fifth part is the velocity gradient in y-direction which is called rate of velocity change i.e. rate of shear strain. The second part is in a form of convolution integral that shows the effect of the fluid memory during the flow process. This integral has two variable functions of $(t-\xi)^{-\alpha}$ and $\partial^2 p / \partial \xi \partial x$ where the first one is a continuous changing function and second one is a fixed function. This means that $(t-\xi)^{-\alpha}$ is an overlapping function on the other function, $\partial^2 p / \partial \xi \partial x$, in the mathematical point of view. These two functions depend on the space, time, pressure, and a dummy variable.

CASE STUDY

The results of the stress-strain rate based on the model presented in Equation (10) can be obtained by solving this equation. In this paper, we focused on the second part of the formulation that is related to the effect of fluid memory and finding a numerical description for a sample reservoir.

A reservoir of length (L = 5000.0 m), width ($W=100.0$ m) and height, ($H = 50.0$ m) have been considered. The porosity and permeability of the reservoir are 30% and 30md, respectively. The reservoir is completely sealed and produces at a constant rate where the initial pressure is $p_i = 27579028$ pa (4000 psia). The fluid is assumed to be API 28.8 gravity crude oil with the properties $c = 1.2473 \times 10^{-9}$ $1/pa$, $\mu_0 = 87.4 \times 10^{-3}$ Pa-s at 298°K. The initial production rate is $q_i = 8.4 \times 10^{-9}$ m^3/sec and the initial fluid velocity in the formation is $u_i = 1.217 \times 10^{-5}$ m/sec. The fractional order of differentiation, $\alpha = 0.2 - 0.8$, $\Delta x = 100.0$ m; and $\Delta t = 7.2 \times 10^4$ sec have also been considered. The computations are carried out for $Time = 10, 50$ and 100 months and at a distance of $x = 1500, 3000$ and 4500 meters from the wellbore. In solving this convolution integral with memory, trapezoidal method is used. All computation is carried out by Matlab 6.5.

Pressure Distribution

To solve the convolution integral in Equation (10), it is necessary to obtain the pressure distribution along the reservoir. The pressure distribution is assumed to be modeled through the diffusivity equation in porous media. This equation has been derived by combining the continuity equation with the Darcy's law as the momentum equation.

$$\frac{\partial^2 p}{\partial x^2} = \frac{\phi \mu_0 c}{k} \frac{\partial p}{\partial t}$$

(11)

To solve this equation, the following dimensionless parameters are considered.

$$x^* = \frac{x}{L}, \; t^* = \frac{u_i t}{L}, \; p^* = \frac{p}{p_i}, \text{ and } q^* = \frac{q}{q_i}$$

(12)

Using the dimensional parameters from Equation (12), Equation (11) can be written in dimensionless form as

$$\frac{\partial^2 p^*}{\partial x^{*2}} = \frac{L \phi \mu_0 c u_i}{k} \frac{\partial p^*}{\partial t^*}$$

(13)

It is defined that $a_2 = \dfrac{L \phi \mu_0 c u_i}{k}$ in Equation (13) and therefore, it may be written that

$$\frac{\partial p^*}{\partial t^*} = \frac{1}{a_2} \frac{\partial^2 p^*}{\partial x^{*2}}.$$

(14)

The initial condition is $p^*(x^*, 0) = 1$. The boundary condition is $\partial p^*(1, t^*)/\partial x^* = 0$ at the outer boundary and the boundary condition at the wellbore is $q^* = c_1 \partial p^*/\partial x^*$, where

$c_1 = -kA_{yz} p_i / \mu L q_i$. The implicit scheme is applied to solve Equation (14) with finite difference method. The discretized form of Equation (14) is

$$p_i^{*(n+1)} = p_i^{*n} + \frac{1}{a_2} \frac{\Delta t^*}{\left(\Delta x^*\right)^2} \left[p_{(i+1)}^{*n} - 2p_i^{*n} + p_{(i-1)}^{*n} \right]$$

(15)

where Δt^* is the time step and Δx^* is the grid size. According to James et al. (1993), the solution of Equation (15) will be stable and nonoscillatory for boundary conditions that remain constant with time if $\Delta t^* / a_2 \left(\Delta x^*\right)^2 \le 0.25$ and will be stable only if $0.25 < \Delta t^* / a_2 \left(\Delta x^*\right)^2 \le 0.50$. Equation (15) can be written as

$$p_i^{*(n+1)} = \left(1 - 2a_1\right) p_i^{*n} + a_1 \left(p_{(i+1)}^{*n} + p_{(i-1)}^{*n} \right)$$

(16)

where, $h = \Delta t^* / \left(\Delta x^*\right)^2$ and $a_1 = h / a_2$. This equation can be solved using the specified initial and boundary conditions.

Figure 2 shows the dimensionless pressure variation with dimensionless distance for different dimensionless time. The pressure is increasing with distance from the wellbore towards the outer boundary. In the boundary, the pressure remains almost constant. As time passes, the pressure starts to decrease more rapidly towards the outer boundary and it goes down more comparing with previous time step.

Figure 3 shows the dimensionless pressure variation with dimensionless time for different dimensionless distance. The trend of pressure is decreasing with time towards the boundary of the reservoir. The pressure decline is sharper and more rapid with time from wellbore towards the boundary. This indicates that at the initial stage of the production, the pressure decline is lower than the one encountered in the next production level by the boundary of the reservoir.

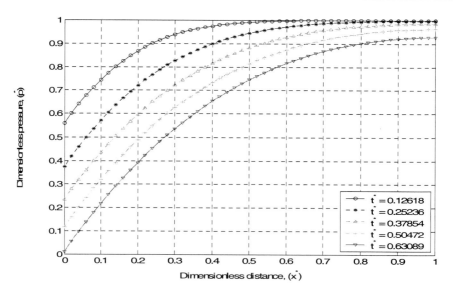

Figure 2. Dimensionless pressure variation with dimensionless distance.

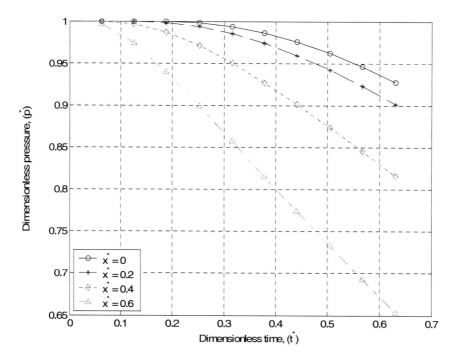

Figure 3. Dimensionless pressure variation with dimensionless time for different dimensionless distance.

Figure 4(a) represents the pressure gradient as a function of time for different distances from wellbore towards the outer boundary of the reservoir. Initially, the pressure gradient is less affected around the wellbore. It rises with time gradually around the wellbore and the slope of the gradient is positive. At the boundary, the pressure gradient is constant. At the wellbore, it approaches zero.

Figure 4(b) shows the pressure gradient in x-direction as a function of distance from the wellbore towards the outer boundary of the reservoir for different time steps. Initially the pressure gradient drop is rapid and higher around the wellbore. It drops very fast and sharply from wellbore toward reservoir boundary. As time passes, pressure gradient decline goes slower with distance and become more stable. This indicates that for larger and larger time, the pressure derivative goes towards its stable condition.

Figure 4(c) displays the pressure gradient change over time in the x direction ($\partial^2 p/\partial x\,\partial t$) as a function of time for different distance values. The $\partial^2 p/\partial x\,\partial t$ change is not very recognizable around the wellbore side throughout the life span of a reservoir. However, this change becomes more important, reportable, and influential at the initial stage of the reservoir production in the position of its outer boundary region and its surroundings.

Figure 4(d) displays the pressure gradient change over time in the x direction ($\partial^2 p/\partial x\,\partial t$) as a function of distance for different time steps. Initially, $\partial^2 p/\partial x\,\partial t$ change is very sharp and affective especially around the wellbore. It turns out to reduce also very quickly with distance towards the outer boundary. However, this pattern changes gradually with time. These variations throughout the reservoir become closer for longer time. The direction of slope change (i.e. from positive to negative) shifts its position from wellbore side towards the outer boundary with time.

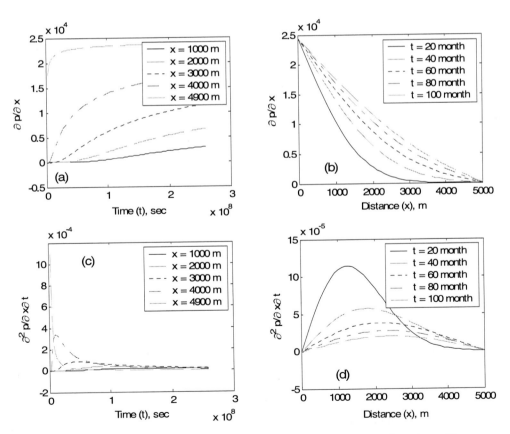

Figure 4. Pressure gradient in x direction for different time steps and distance from the wellbore.

Figure 5(a) shows the pressure change over time as a function of time for different distances from the wellbore. Throughout the reservoir life, the pressure change over time is not very affective especially around the wellbore. However, this pattern changes gradually towards the outer boundary of the reservoir with time. The direction of the curve slope and shape change is very fast during the initial stage throughout the reservoir. These variations throughout the reservoir become closer for longer time period.

Figure 5(b) shows the pressure change over time as a function of distance from the wellbore towards the boundary of the reservoir for different time steps. The direction pattern of the pressure change over time is characterized by an *S*-shape. At the initial stage of production, the shape and the direction of the pressure change over time illustrate rapid and sharp variation around the wellbore. However, this pattern changes gradually with time and advances to become constant towards the outer boundary of the reservoir. The slope of the curve decreases gradually comparing with previous time step. This advancement indicates that the pressure change over time reaches its steady state for larger and larger time. Therefore, the reservoir behavior is more stable in the long term than in the short term.

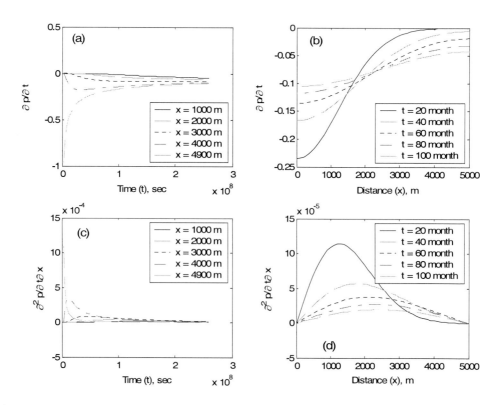

Figure 5. Pressure change over time and distance in the x direction for different time steps and distance from the wellbore.

Figure 5(c) displays the pressure gradient change over time in the *x* direction ($\partial^2 p / \partial t\, \partial x$) as a function of time for different distance values. Figure 5(d) displays the pressure gradient change over time in the x direction ($\partial^2 p/\partial t\, \partial x$) as a function of distance for different time steps. Both figures are same as figures 4(c) and 4(d).

RESULTS AND DISCUSSION

The stress-strain relationship is given in Equation 10. The effect of the memory of the fluid flow is distributed in the second and third parts of it as we describe it already. Here, we have the fluid memory parameters with pressure distribution in space and time. The ratio of the pseudopermeability of the medium with memory to fluid viscosity (η) is considered as a constant. So, the effect of memory due to this parameter is constant with pressure distribution in the formation. It does not change with space and time. The most influential parameter for the effect of memory is the fractional order of differentiation (α) which is related to pressure distribution, space and time. This is in the second part of Equation 10 and identified by I_{mi} as;

$$I_{mi} = \left\{ \frac{\int_0^t (t-\xi)^{-\alpha} \left[\partial^2 p/\partial\xi\, \partial x\right] d\xi}{\Gamma(1-\alpha)} \right\}^{0.5}$$

(17)

In Equation 11, the fractional derivative α exists on both numerator and denominator. It ranges from 0 to 1. The denominator, $\left[\Gamma\left(1-\alpha\right)\right]^{0.5}$, varies as a function of α as presented in Figure 6. It changes as follows: $1 \leq \left[\Gamma\left(1-\alpha\right)\right]^{0.5} < \infty$.

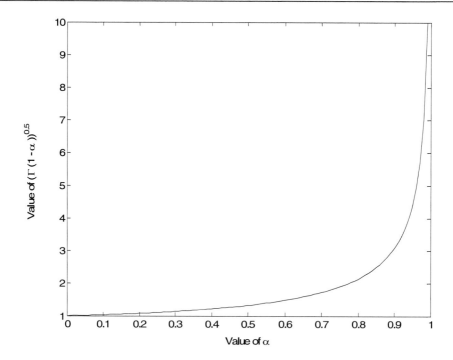

Figure 6. Variation of Gamma part of memory of Equation (10) over α .

DEPENDENCE ON TIME

At Different Distance from the Wellbore

The effect of the memory is given in Equation 11 and identifies By I_{mi}. This is a function of time and space. We studied the variation of it with respect to time and space for the sample reservoir that we explained already. Duration of 100 months is taken into account to find the effect of value of memory at different distances from the wellbore.

Figure 7 shows the change of I_{mi} for different α values for duration of 100 months at a distance of 1500 m from the wellbore. The value of I_{mi} is varied between 0.0032 and 68.2569 for $\alpha = 0$. Initially, it remains same for the production period of 3 months and increases faster with a small increase of time. Finally its increase rate slows down for longer period of time. The trend is almost the same for the other values of α but the magnitude of I_{mi} is decreased up to $\alpha = 0.4$ and then increased as depicted in the figure. In Figure 7, the value of I_{mi} changes from 0.0007 to 10.3108 for $\alpha = 0.2$. The magnitudes of I_{mi} are much less than that acquired earlier. This indicates the existence of fluid memory in nature because the consideration of α affects the resulting values of the memory part. However, the figure indicates a strong dependency of the value of I_{mi} to the variation of α that affected the stress-strain relationship. At the beginning of the production, the memory part decreases to a value of 0.0006 and remains constant for the period of 3 months. After this, it starts to increase faster with the increase of time. The increasing rate of the memory part slows down after 19 months of production which continues until the end of the reservoir production life.

Figure 7 shows also that, for $\alpha = 0.4$, I_{mi} changes from 0.0002 to 2.8260 which are lower than the latter values. As production begins, I_{mi} decreases up to 0.0001 after 2 months of production and increases gradually up to 0.0651 for a period of 8 months. After this time, it increases faster with the increase of production time. This trend remains up to 2.8555 for the period of 84 months and remains same up to 86 months and then starts to decrease with time. For $\alpha = 0.6$, I_{mi} changes from 0.0002 to 7.7677 which are greater than those corresponding to $\alpha = 0.4$. The memory part decreases very fast up to 0 for a small production period of 2 months whereas it remains same in the previous two α values.

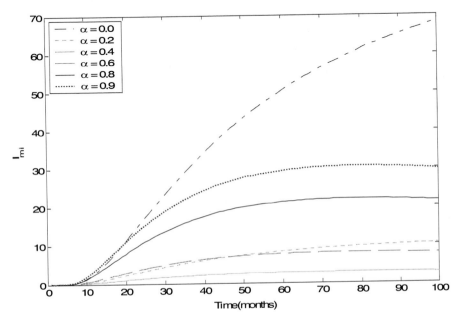

Figure 7. change of memory part over time as a function of different α values at a distance of 1500 m from the wellbore.

These changes show the chaotic and unpredictable fluid flow within the reservoir due to its inherent fluid memory. After this point, it increases very fast with increase of production time up to a level of 6.9868 for 47 months and then the increasing trend slows down for 7.9595 at 79 months. This trend starts to decrease with time beyond this production period. In addition, the magnitude of I_{mi} varies between 0.0006 and 21.4586, which is plotted for $\alpha = 0.8$. These values are greater than those obtained for $\alpha = 0.6$. The memory part decreases very fast up to 0 for a small production period of 2 months. After this point, it increases faster with increase of production time up to a level of 20.9505 for 58 months and then the increasing trend slows down for 21.9887 at 79 months. This trend starts to decrease with time beyond this production time. For $\alpha = 0.9$, I_{mi} varies between 0.0009 and 29.7228 (see Figure 7). The magnitude of I_{mi} is greater than that of the previous α. This case displays a decreasing trend, which is similar to that obtained for $\alpha = 0.8$. However, this trend increases faster with increase of production time up to a level of 29.9437 for 66 months and then the increasing trend slows down for 30.4571 at 80 months.

Figure 8 shows the change of I_{mi} for different α values during a production period of 100 months at a distance of 3000 m from the wellbore. For $\alpha = 0$, I_{mi} changes from 1.9049 to 107.2981. Initially, it increases gradually up to 3.5120 for the production period of 4 months and then increases rapidly with a small increase of time. It gradually tends to reach its steady state in the long term. For $\alpha = 0.2$, I_{mi} varies from 0.4309 to 15.5123. This is much less than that encountered in the previous case but much greater than the one shown in Figure 7. This indicates that memory effects increase with the increase of time and distance. I_{mi} increases gradually from the beginning of production to 4 months and then increases rapidly with a small increase of time. It tries to reach its steady state in the long term. For $\alpha = 0.4$, I_{mi} changes from 0.1022 to 3.4149. This magnitude is less than the previous one obtained for $\alpha = 0.2$, but much greater than the one in the case of Figure 7. I_{mi} decreases very fast up to 0.0663 after 2 months of production and increases also very fast with increase of product on time. This trend remains the same up to 3.6038 for the period of 63 months and then starts to decrease with time. For $\alpha = 0.6$, I_{mi} variation of 0.1276 to 8.5740 is greater than those changes observed for $\alpha = 0.4$. This is due to the chaotic and unpredictable behavior of fluid which is explained by fluid memory. Here, I_{mi} decreases very fast to 0 for a small production period of 2 months. After this point, it increases very fast with increase of time of production. This trend remains the same up to 9.4141 for the period of 51 months and then starts to decrease with time. For $\alpha = 0.8$, I_{mi} varies from 0.3488 to 23.6813. These values are greater than those obtained for $\alpha = 0.6$. I_{mi} decreases very fast up to 0 for 2 months and increases very fast with increase of time of production. This trend remains the same up to 26.0396 for the period of 50 months and then starts to decrease with production time. For $\alpha = 0.9$. I_{mi} varies from 0.4834 to 32.8016 which are greater than those for $\alpha = 0.8$ of Figure 8 and $\alpha = 0.9$ of Figure 7. I_{mi} decreases very fast to 0 for 2 months and increases very fast with increase of production time. This trend remains the same up to 36.0949 for the period of 50 months and then starts to decrease with production time.

Figure 9 shows the variation of I_{mi} for different α values for a production duration of 100 months at a distance of 4500 m from the wellbore. For $\alpha = 0$, I_{mi} varies from 65.7528 to 153.5833. Initially, it increases rapidly with a small increase of time and gradually tries to reach steady state with time. For $\alpha = 0.2$, I_{mi} changes from 15.4573 to 20.9319. These values are much less than those given in the previous case, but much greater than those of Figure 8.

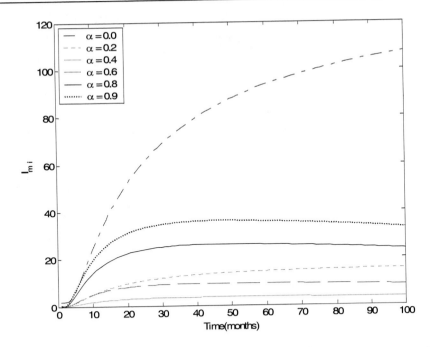

Figure 8. change of memory part over time as a function of different α values at a distance of 3000 m from the wellbore.

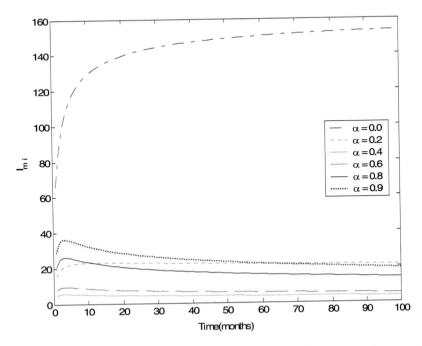

Figure 9. change of memory part over time as a function of different α values at a distance of 4500 m from the wellbore.

This indicates that memory effects increase with the increase of distance from the wellbore and shows also a strong influence of fluid memory. I_{mi} increases very fast at the beginning of production. This goes to its peak value of 22.9698 after 13 months of the production and then starts to reduce with time. The decreasing trend expedite towards the end of production life. For $\alpha = 0.4$, I_{mi} varies from 4.2367 to 3.1239 which are much less than those obtained for $\alpha = 0.2$. Here I_{mi} value is much higher at the beginning of production and less at the end period of production comparing with the case of 3000 m shown in Figure 8. However, I_{mi} increases very fast up to 5.0823 after 4 months of production and goes down also very fast with increase of production time. For $\alpha = 0.6$, I_{mi} ranges from 7.5237 to 4.9563 which are greater than those acquired for $\alpha = 0.4$. This behavior is similar to the previous one of Figure 8. Here, I_{mi} increases very fast up to 9.2729 after 4 months of production and goes down also very fast with increase of time of production. Figure 9 shows also the plot for $\alpha = 0.8$ where I_{mi} varies from 20.7022 to 13.6660. The magnitudes are greater than the previous $\alpha = 0.6$. I_{mi} increases very fast up to 26.1875 for a shorter period of production time, 3 months, and after that time, it goes down gradually with increase of time of production. In the same figure, the variation of I_{mi} is shown for $\alpha = 0.9$ which varies from 28.6935 to 18.9291. The trend and magnitudes of I_{mi} are greater than those of $\alpha = 0.8$ which increases very fast up to 36.2959.

At Different Reservoir Production Life

As indicated earlier in this study, memory is a property of a fluid which represents the history of a fluid. So, time is the most important parameter here. Moreover, I_{mi} is a function of time whereas the pressure gradient changes over time in the x direction ($\partial^2 p / \partial t\, \partial x$) is also a function of time. Therefore, the reservoir life is considered to analyze memory of fluid.

The convolution integral part, $I_{ci} = \left\{ \int_0^t (t - \xi)^{-\alpha} \left[\partial^2 p / \partial \xi\, \partial x \right] d\xi \right\}^{0.5}$ i.e. only the numerator is considered in computation because there is no time dependency in the denominator. In this study, reservoir production life is considered as 10, 50 and 100 months.

Table 1 shows the different data of the convolution integral part at different reservoir distance from the wellbore for the reservoir life of 10 months. The data shows that when distance from the wellbore increases, the variation of I_{ci} increases for a particular value of α. I_{ci} also increases with the increases of α. This indicates that memory effect is more dominant as the distance increase from the wellbore. Figure 10 shows the variation of I_{ci} as a function of α for all the data showed in Table 1. The shape and pattern except the magnitude of I_{ci} of the figures are identical for different distances. The common trend of all figures are initially decreasing and then after approximately $\alpha = 0.5$, I_{ci} starts to increase with the increase of α value.

Table 2 shows the different data of I_{ci} at different reservoir distance from the wellbore for the reservoir life of 50 months. The data shows that when distance from the wellbore increases, the variation of I_{ci} increases for a particular value of α up to $\alpha = 0.2$ and beyond this, it decreases with α. This indicates a chaotic behavior of α values.

Table 1. Numerical values of Ici for different α values in different distances for a reservoir life of 10 months

Value of α	I_{ci} for 10 months		
	x = 1500 m	x = 3000 m	x = 4500 m
0.0	1.3713	24.1142	128.7294
0.2	0.3369	5.3832	24.7316
0.4	0.1987	2.2587	5.7501
0.6	0.7158	7.5266	12.5635
0.8	2.8485	29.9444	49.8758
0.9	5.6840	59.75116	99.5226

Table 2. Numerical values of I_{ci} for different α values in different distances for a reservoir life of 50 months

Value of α	I_{ci} for 50 months		
	x = 1500 m	x = 3000 m	x = 4500 m
0.0	42.9970	87.3485	148.9504
0.2	7.9002	15.0594	23.6703
0.4	3.0771	4.4089	4.3923
0.6	10.7237	14.0209	8.9366
0.8	42.6821	55.7932	35.4989
0.9	85.1688	111.3310	70.8352

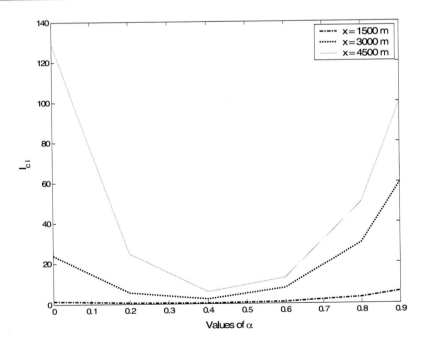

Figure 10. change of memory part over α values for the reservoir life of 10 months at different distance from the wellbore.

Figure 11 shows the variation of I_{ci} as a function of α for all the data showed in Table 2. This is same as Figure 10 except the magnitude of I_{ci}.

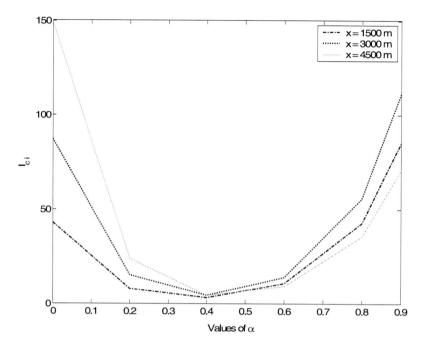

Figure 11. change of memory part over α values for the reservoir life of 50 months at different distance from the wellbore.

Table 3 shows the different data of I_{ci} at different reservoir distance from the wellbore for the reservoir life of 100 months. The data shows that when distance from the wellbore increases, the variation of I_{ci} increases for a particular value of α up to $\alpha = 0.2$ and beyond this, it decreases with α. Figure 12 shows the variation of I_{ci} as a function of α for all the data showed in Table 3. This is same as Figure 11 except the magnitude of I_{ci}.

Table 3. Numerical values of I_{ci} for different α values in different distances for a reservoir life of 100 months

Value of α	I_{ci} for 100 months		
	x = 1500 m	x = 3000 m	x = 4500 m
0.0	68.2569	107.2981	153.5833
0.2	11.1253	16.7377	22.5854
0.4	3.4487	4.1673	3.8121
0.6	11.5688	12.7697	7.3816
0.8	45.9777	50.7402	29.2812
0.9	91.6770	101.1732	58.385

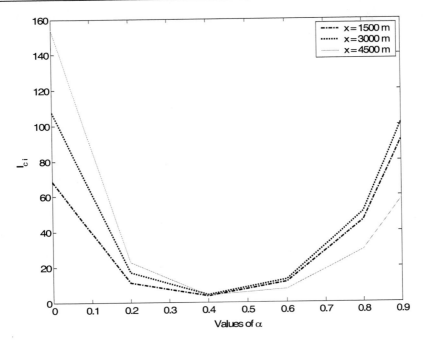

Figure 12. change of memory part over α values for the reservoir life of 100 months at different distance from the wellbore.

CONCLUSION

A comprehensive model is introduced for stress-strain rate involving the fluid memory effect. It takes into account the influence of temperature, pressure, and surface tension variations. The effect of temperature on viscosity is applied by using the Arrhenius model. This model expresses the variation of temperature with an exponential function. The pressure variation is applied using the Darcy's law. This law is an experimental relation that is extensively used in the fluid flow in porous media. The effect of surface tension is included through the Marangoni number. This model gives the surface tension gradients along a gas–oil interface which provokes strong convective activity, called Marangoni effect. The fluid memory is incorporated with application of α variation.

The memory effect causes a nonlinear relation between the stress and strain rate. The part of the stress-strain formulation related to the memory is taken into account and we obtain the variation of it with time and distance for $0 \leq \alpha < 1$. The memory part that is denoted by I_{mi} in the paper has a mild variation for a time duration of 100 months near the wellbore. The variation of I_{mi} is increased as the distance increase from the wellbore. It contained a peak value and then the value of I_{mi} increased mildly if the value of α is less than 0.5. The variation of I_{mi} is increased with increasing the value of α after that value of α. The distance from the wellbore also affects the fluid memory part in the developed model. The memory effects increase with the increase of distance from the wellbore at the beginning of production or for a shorter reservoir life. When the life of the reservoir increases, the memory effects increase for lower values of α whereas they decrease for higher values of α. The variation of distance and time defines a chaotic behavior with increasing and decreasing trends due to α variation in I_{mi} and I_{CS} magnitudes, which gives a strong indication of memory effect. This paper reveals that the memory mechanism helps in interpreting the reservoir phenomenology in addition to other parameters such as matrix heterogeneity, anisotropy, and inelasticity.

ACKNOWLEDGEMENT

The authors would like to thank the Atlantic Canada Opportunities Agency (ACOA) for funding this project under the Atlantic Innovation Fund (AIF). The first author would also like to thank Natural Sciences and Engineering Research Council of Canada (NSERC) for funding.

REFERENCES

Abel, M. S., Khan, S. K., Prasad, K.V. 2002. Study of visco-elastic fluid flow and heat transfer over a stretching sheet with variable viscosity. *International Journal of Non-Linear Mechanics*, 37: 81-88.

Avramov, I. 2005. Viscosity in disordered media. *Journal of Non-Crystalline Solids*, 351: 3163–3173.

Brenner, H. 2005. Navier–Stokes revisited. *Physica* A 349, 60–132.

Broszeit, J. 1997. Finite-element simulation of circulating steady flow for fluids of the memory-integral type: flow in a single-screw extruder. *Journal of Non-Newtonian Fluid Mechanics,* May, Vol –70, Issues – 1-2, 35-58.

Caputo, M. 1999. Diffusion of fluids in porous media with memory. *Geothermics*, 23, 113-130.

Ciarletta, M. and Scarpetta, E. (1989). Minimum Problems in the Dynamics of Viscous Fluids with Memory. *International Journal of Engineering Science*, Vol – 27, No – 12, 1563-1567.

D'Aubeterre, A., Silva, R. Da, and Aguilera, M.E., 2005. Experimental Study on Marangoni Effect Induced by Heat and Mass Transfer. *International Communications in Heat and Mass Transfer* 32: 677-684.

Das, K.S., and Bhattacharjee, J.K. 1999. Marangoni Attractors. *Physica* A, 270: 173-181.

Eringen, A. C. 1991. Memory Dependent Orientable Nonlocal Micropolar Fluids. *International Journal of Engineering Science*, Vol – 29, No – 12, 1515-1529.

Gan, Q., Xue, M., and Rooney, D. 2006. A study of fluid properties and microfiltration characteristics of room temperature ionic liquids $[C_{10\text{-min}}][NT_{f2}]$ and $N_{8881}[NT_{f2}]$ and their polar solvent mixtures. *Separation and Purification Technology*, in press.

Haferl, S., and Poulikakos, D. 2003. Experimental Investigation of the transient impact fluid dynamics and solidification of a molten micro droplet pile-up. *International Journal of Heat and Mass Transfer,* 46: 535-550.

Haminiuk, C.W.I., Sierakowski, M.R., Vidal, J.R.M.B., and Masson, M.L. 2006. *Influence of temperature on the rheological behavior of whole araca' pulp* (Psidium cattleianum sabine). LWT, 39: 426–430.

Hossain, M. E., Mousavizadegan, S. H. and Islam, M. R. 2008. Rock and Fluid Temperature Changes during Thermal Operations in EOR Processes. *Journal Nature Science and Sustainable Technology*, in press.

Hossain, M.E. and Islam, M.R. 2006. Fluid properties with memory – A critical Review and some additions. Proceeding of 36th *Int. Conf. Comp. Ind. Eng.*, June 20-23, Taipei, Taiwan, Paper Number: CIE – 00778.

Hu, X., and Cushman, H., 1994. Non equilibrium statistical mechanical derivation of a nonlocal Darcy's law for unsaturated/saturated flow. *Stochastic Hydrology and Hydraulics*, 8, 109-116.

James, M.L, Smith, G.M., and Wolford, J.C. 1993. *Applied Numerical Methods for Digital Computation*. HarperCollins College Publishers, Fourth edition, New York, NY 10022, USA, p. 614.

Jossi, J.A., Stiel, L.I., and Thodos, G., 1962. The viscosity of pure substance in the dense gaseous and liquid phases. *A.I.Ch.E. Journal*, 8, 59.

Kozak, R., Saghir, M.Z., and viviani, A. 2004. Marangoni Convection in a liquid layer overlying a Porous Layer with Evaporation at the free Surface. *Acta Astronautica*, 55: 189-197

Li, H.Z., Frank, X., Funfschilling, D., Mouline, Y. 2001. Towards the Understanding of Bubble interactions and Coalescence in non-Newtonian Fluids: a cognitive approach. *Chemical Engineering Science*, 56, 6419-6425.

Lyford, P., Pratt, H., Greiser, F., Shallcross, D. 1998a. The Marangoni Effect and Enhanced Oil Recovery Part 1. Porous Media Studies. *Can. J. Chem. Eng.*, 76, April, 167-174.

Lyford, P., Pratt, H., Greiser, F., Shallcross, D. 1998b. The Marangoni Effect and Enhanced Oil Recovery Part 2. Interfacial Tension and Drop Instability. *Can. J. Chem. Eng.*, 76, April, 175-182.

Nibbi, R. 1994. Some properties for Viscous Fluids with Memory. *International Journal of Engineering Science*, Vol – 32, No – 6, 1029-1036.

Recondo, M.P., Elizalde, B.E., and Buera, M.P. 2005. Modeling temperature dependence of honey viscosity and of related supersaturated model carbohydrate systems. *Journal of Food Engineering*, in press.

Bragard, J. and Velarde, M. 1998. Benard –Marangoni convection: Planforms and related theoretical predictions. *J. of Fluid Mechanics*, Vol – 368, August 10, 165-195.

Zuritz, C.A., Puntes, E. M., Mathey, H.H., Pe´rez, E.H., Gasco´n, A., Rubio, L.A., Carullo, C.A., Chernikoff, R.E., and Cabez, M.S. 2005. Density, viscosity and coefficient of thermal expansion of clear grape juice at different soluble solid concentrations and temperatures. *Journal of Food Engineering*, 71: 143–149.

INDEX

E

F

G

H

Q

R